ADVANCES IN SPACE RESEARCH

The Official Journal of the Committee on Space Research (COSPAR)
A Scientific Committee of the International Council of Scientific Unions (ICSU)

VOLUME 20, NUMBER 7

PROBLEMS OF SPACE SCIENCE EDUCATION AND THE ROLE OF TEACHERS

Symposium PSRDC1

Sponsors

COMMITTEE ON SCIENCE AND TECHNOLOGY IN DEVELOPING COUNTRIES/INTERNATIONAL BIOSCIENCES NETWORKS (COSTED-IBN)
COMMITTEE ON SPACE RESEARCH (COSPAR)
INTERNATIONAL UNION OF BIOLOGICAL SCIENCES (IUBS)
SCIENTIFIC COMMISSIONS A AND F
THIRD WORLD ACADEMY OF SCIENCES (TWAS)
UN OFFICE FOR OUTER SPACE AFFAIRS

Program Committee

K. Kasturirangan, India
R. S. Young, U.S.A.
S. C. Chakravarty, India
J.-L. Fellous, France
J. Rummel, U.S.A.
J. Narlikar, India
Y. X. Yan, China
A. A. Abiodun, Austria
M. H. A. Hassan, Italy
R. R. Daniel, India
T. Owen, U.S.A.
J. Ortner, Austria

PROBLEMS OF SPACE SCIENCE EDUCATION AND THE ROLE OF TEACHERS

Proceedings of the PSRDC1 Symposium of COSPAR Panel on Space Research in Developing Countries which was held during the Thirty-first COSPAR Scientific Assembly, Birmingham, U.K., 14–21 July 1996

Edited by

S. C. CHAKRAVARTY

Indian Space Research Organisation, Antariksh Bhavan, New BEL Road, Bangalore, 560094, India

J.-L. FELLOUS

MEDIAS-FRANCE, CNES, 18 Ave Edouard Belin, Bpi 2101, Toulouse 31401 Cedex 4, France

K. KASTURIRANGAN

Indian Space Research Organisation, Antariksh Bhavan, New BEL Road, Bangalore, 560094, India

and

M. J. RYCROFT

Parc d'Innovation, Boulevard Gonthier d'Andernach, Illkirch 67400, France

Published for

THE COMMITTEE ON SPACE RESEARCH

PERGAMON

U.K.	Elsevier Science Ltd, The Boulevard, Langford Lane, Kidlington, Oxford OX5 1GB, U.K.
U.S.A.	Elsevier Science Inc., 660 White Plains Road, Tarrytown, New York 10591-5153, U.S.A.
JAPAN	Elsevier Science Japan, Tsunashima Building Annex, 3-20-12 Yushima, Bunkyo-ku, Tokyo 113, Japan

---

First edition 1997

ISBN 0-08-043304-9

In order to make this volume available as economically and as rapidly as possible the author's typescript has been reproduced in its original form. This method unfortunately has its typographical limitations but it is hoped that they in no way distract the reader.

Transferred to digital printing 2005

CONTENTS

Pergamon

Adv. Space Res. Vol. 20, No. 7, pp. 1323–1324, 1997

Printed in Great Britain
0273-1177/97 $17.00 + 0.00

PII: S0273–1177(97)00723–0

PREFACE

Considering that space science has been a major source of inspiration in various scientific endeavours for national development and taking into account the ground realities that many developing countries are yet to benefit from a full opportunity to share or participate in the current excitement of space sciences as a means of improving the quality of human resources, COSPAR Panel on Space Research in Developing Countries (PSRDC) organised a meeting to discuss the related issues during the 31st COSPAR Scientific Assembly held in Birmingham. The main focus of the discussions, cosponsored by the COSPAR Commissions A & F, the United Nations, the Committee on Science & Technology in Developing Countries (COSTED) and the Third World Academy of Sciences (TWAS), were centred around the need and the problems of introducing space science education in various levels of academic curricula and the role of motivated and qualified teachers in this endeavour. The main scientific objective of the meeting, attended by about 50 participants with 22 invited and contributed presentations, was to identify and understand various problems experienced to introduce and implement space science education at academic and professional levels and to improve public awareness and education about the excitement and benefits of space research. The deliberations of the scientific presentations touched upon a number of topics dealing with the variants of structured syllabi, spectrum and usefulness of different approaches and methodologies to impart space science education, specialised training of teachers and their involvement in space science research projects sponsored by the space agencies. One of the main goals of this meeting was to generate a set of recommendations to improve the understanding and devise strategies related to effective introduction of space science education in academic curricula without interfering with the training programmes in other basic science subjects.

Delivering his inaugural address, Prof. G. Haerendel, President COSPAR mentioned how COSPAR is committed to provide wholehearted support to the recommendations of PSRDC on promotion of space research in developing countries. He highlighted the need to utilise the global resource of information and data related to space science and applications. In his keynote address, Prof. Sir Hermann Bondi emphasised the inspiration derived by mankind from space research and the need to pursue basic space science studies in schools and colleges. He stressed that the spectacular success story of space science research is perhaps the best example in human co-operation and understanding of the highest order.

The meeting had four technical sessions dealing with (a) relevance and challenges of space science education: current efforts and future plans in this direction, (b) approach, curricula, means and methodology of imparting space science education at various academic levels and for building public awareness, (c) refresher training programme for space science teachers and encouraging their participation in space research activities and (d) specific case studies of the current and future space science education programmes in developing countries of Asia, Africa and Latin America. In their presentations, the authors emphasised the need for training of teachers, establishment of UN sponsored centres for space science and technology education at the regional level, use of computer and communication networks in aid of space science education, weather/remote sensing data reception, processing and analysis particularly meant to study the global change phenomenon.

The scientific and technical presentations were followed by in-depth discussions leading to the formulation of a set of recommendations which includes (a) space science should be taught as part of the academic and professional courses, primarily to enhance general literacy in space science and technology, (b) the vast global resource of space based data and information needs to be utilised by the scientists from developing countries, to augment their in-house education/training programmes, (c) innovative audio-visual tools, models, kits etc., should be used along with effective electronic communication facilities to impart space science education in schools, colleges and universities, (d) space science education is to be organised and strengthened with international co-operation for access to information and data, availability of text books/reading material, and teaching aids.

While evaluating the papers appearing in this issue, particular effort has been made to maintain the originality of the manuscripts, continuity and coherence of thoughts.. The collection in this issue, when made available to the developing countries, will be one more step in the direction of creating a broader global base for space science education, research and applications. PSRDC would do its part to encourage and support such activities and also urge various national and international agencies to encourage and sponsor space science education and related projects in the developing countries. We are sincerely thankful to Professor R. R. Daniel and Professor M. J. Rycroft, the two main referees for their painstaking job which resulted in the modified manuscripts for publication. We would also like to thank all the co-sponsors, the members of the organising committee and the editors for their cooperation.

K. Kasturirangan
(Chairman, PSRDC)

Pergamon

Adv. Space Res. Vol. 20, No. 7, pp. 1325–1326, 1997

Printed in Great Britain
0273-1177/97 $17.00 + 0.00

PII: S0273–1177(97)00724–2

PRESIDENTIAL ADDRESS

On behalf of the COSPAR Council and the Bureau, it is my great pleasure to extend to you greetings and best wishes for a successful meeting. We welcome very much the activities of PSRDC, in particular, the subject matter of this meeting. It is completely in line with Decision No. 2/94 approved in 1994 by the COSPAR Council at its 30th Scientific Assembly held in Hamburg. I quote from this decision as follows:

"...wider participation of the global scientific community is of importance to studies relevant to global and regional change .... COSPAR could play a useful role in promoting wider participation and in enhancing the scientific utility of data gathered from outer space; COSPAR resolves: ... that collaboration with relevant international programs and organizations should ... include aspects relating to education and training, data accessibility, information assistance and provision of facilities for scientific work."

Furthermore, in the decision COSPAR urges its national members to encourage and support related programs and activities. Why should COSPAR be interested in such educational aspects? The answer is contained in our Charter: "... to promote on an international level scientific research in space, with emphasis on the exchange of results, information and opinions, and to provide a forum, open to all scientists for the discussion of problems that may affect scientific space research". Promotion of space research in countries where it hardly exists or does not exist at all, requires in the first place, education about (i) problems and the ways of contributing to their solution through space techniques, (ii) handling and interpreting space data and (iii) building instrumentation for space applications. The level of preparedness for such educational programs may vary greatly, but the intellectual curiosity exists widely around the globe, although, often in great isolation. How to overcome this isolation, may be regarded as one of the foremost tasks of such education programs.

COSPAR responds actively to Decision 2/94, and this fact is supported by numerous activities of our Committee including the COSPAR/IAF Symposia held during the annual meetings of the Scientific and Technical Sub-committee of COPUOS in Vienna. In particular, the Symposia held in February 1995 and 1996 were of great relevance to Decision 2/94. The subject of the 1995 symposium, "Application of space technology for education with particular emphasis on its use in developing countries," speaks for itself. This year's theme of the use of micro and small satellites was again directed toward developing countries and contained a large educational element.

In addition to organizing such symposia and giving financial support to participants from less developed countries (LDCs), COSPAR is initiating a program to bring together those who have an excess of space data and those who badly want to work with them but are unable to do so because of their isolation. The idea is that COSPAR will be instrumental in organizing regional workshops in which experienced researchers from space faring nations equipped with data tapes and software packages meet with LDC "students" and train them in the use of such data. It is anticipated that long lasting collaboration will grow out of these workshops. After approval by its Council, COSPAR will propose this initiative to ICSU in order to obtain financial support.

I would like to add a word based on a personal experience with such regional workshops. Some time in the 1980s, we organized a South American Radar School in Jicamarca, Peru. It was well-attended by "students" from several South American nations, and the teaching by the local, North American and European scientists was of very high level. This event provided enormous encouragement to ionospheric research in South America and resulted in the entry of a few gifted young people to this challenging field of research. In this case the availability of a powerful research facility, the ionospheric incoherent scatter radar at Jicamarca proved to be very useful for accelerated progress. What COSPAR intends to propose goes beyond what I have just described. We aim to exploit intellectual resources lacking proper facilities and even the basic means of data transfer and processing.

Let me finish with a general thought about education. I believe that a most important element in developing technical and other expertise in any field is to follow the advice formulated in the title of a famous essay, "About the gradual perfection of thoughts while talking", by the German poet Friedrich Schiller. One learns about space and various related scientific issues by doing work by once own hand, eg., by analyzing the existing data and gradually making unique contributions to the field. Education by doing and not only by listening to lectures at conferences – this is the new goal of, and challenge for COSPAR.

I wish you a most fruitful workshop.

G.Haerendel
(President, COSPAR)

Pergamon

Adv. Space Res. Vol. 20, No. 7, pp. 1327–1328, 1997

Printed in Great Britain
0273-1177/97 $17.00 + 0.00

PII: S0273–1177(97)00725–4

KEYNOTE ADDRESS ON SPACE SCIENCE EDUCATION

In any educational programme, there needs to be a careful selection of topics. For inevitably teaching time is limited and there is so very much one would like one's students to learn. Which features of space science should be selected?

To me, the human interest comes first: Space science requires human co-operation of a very high order. Specialists in many different disciplines have to work together intensively and over long periods for any scientific result to be obtained. International links have to be established and nurtured. People have to get to know each other and learn to rely on each other. Above all, excellent communication is the essence and must be given the requisite time and resources. Everyone engaged in the task has to learn to collaborate with the others involved in an effective, trusting and preferably enjoyable manner. There is indeed an excellent test at the end of the day: If the experiment in space science works, the intensive human co-operation required has been achieved; if not, then the most likely cause is a failure in human co-operation.

To me, the first task of the teacher is therefore to display space science as a very testing model of what we human beings can achieve across all barriers of language, nationality, religion, specialisms, ideology by working together. Indeed, 'development' must involve educating the whole population into the habit of working together. True 'underdevelopment' is characterised by close co-operation being confined to the family. Space science is a model of what wider and closer co-operation can achieve.

As a corollary the teacher might point out that it is a noble aim to try to participate in this great enterprise, an aim that should encourage students to work hard and with persistence, that should encourage them to learn to involve themselves with others of a different background and, perhaps most significantly, to make a real effort to acquire the ability to communicate concisely with clarity and precision.

In coming next to the aims and achievements of space science it is perhaps too easy to point out their immediate utility (impressive as this is). Like all science it arises from that essential component of the human spirit, our inquisitiveness. Curiosity is certainly nothing to be ashamed of; it is one of the main springs of human endeavour. And what a wealth of interests can be served by space science ! In cosmology it is contributing mightily to our understanding of the universe we exist in. In galactic astronomy it reveals remarkable features of our own and of other galaxies. In stellar astronomy it fosters our knowledge of the enormous variety of stars and other objects. (I like to say that the universe is a well-stocked laboratory with samples of stranger objects than we could imagine). In our own planetary system space science keeps on revealing new and often unexpected features of the other members of the solar system, making us more aware of the features of our planet Earth that make it so habitable. Perhaps this last point needs stressing to make clear to teachers and through them to their students the features of our home in space that need safeguarding against over-exploitation and against population pressures. These worries apply everywhere, but the second one is particularly relevant to developing countries.

It is indeed in the study from space of our Earth and its immediate surroundings (especially its vital relation to the Sun) that the interest and the utility of space science come to the fore. Even where administration and knowledge of land use and soil condition is already good, space based observation has much to add; where such knowledge is more limited (as in many developing countries), the scientific observation and interpretation from space is of immense value. Students might well be taught that while scientific study of the land from space may perhaps be viewed as an optional extra in a well mapped and studied industrialised country, it is a necessity for a developing nation. The only proviso is that before the flood of data space provides can be made useful, the highly skilled staff required for interpretation has to be available and organised.

Climate research is of interest to us all, but the impact of climate variation is immeasurably greater in less developed regions. The vital contribution space science makes to the understanding of climate and its vagaries needs to be taught.

But perhaps I end by stressing that the very term 'space science' is instructive, for it relates to a disparate group of disciplines held together only by the technology of space. It is an excellent lesson for those (perhaps more common in certain parts of the world) who view technology as secondary and less admirable than science. This is nonsense. Science advances primarily through its experiments, which themselves advance through the progress of technology, just as new insights in science help to improve technology. Science and technology are interdependent; to regard one as primary and the other as secondary makes no sense at all. Space science is a splendid teaching example to bring this point home.

Hermann Bondi
(Churchill College, Cambridge,UK)

Pergamon

Adv. Space Res. Vol. 20, No. 7, pp. 1329–1333, 1997

Printed in Great Britain
0273-1177/97 $17.00 + 0.00

PII: S0273–1177(97)00726–6

RELEVANCE AND CHALLENGES OF SPACE SCIENCE EDUCATION IN DEVELOPING COUNTRIES

K. Kasturirangan

Indian Space Research Organisation, Antariksh Bhavan, New BEL Road, Bangalore, 560094, India

ABSTRACT

Development in space research activities during the past decades has been responsible to bring about revolutionary concepts and advances in long distance reliable communications, remote sensing of earth's resources and environment, planetary science, astronomy and cosmology. The ever-unfolding future prospects and potentials of the field, which are to be exploited for the well-being of the human race, pose many challenges including educating the masses. The configuration and content of imparting such education at different academic levels vary with the overall promotion of space research activities of a country. It is recognised that no single system of space science education can perhaps be applied uniformly to all countries as local variables are to be considered based on interest, infrastructure, practical experience and availability of trained manpower. The prevailing situation of science education in general and space science education in particular with reference to developing countries is discussed along with suggestions on the necessary means to improve the education programme in this frontier area of great intellectual possibilities.

INTRODUCTION

The launching and orbiting of the first artificial object in space ushered in a new era of human intellectual involvement to unravel the mysteries of the universe. With the evolution and utilization of this new technological tool, and within a short span of time, humankind has been the witness to several exciting developments - the landing of man on the moon, the perfect touchdown of space shuttles, vivid images of the surface of mars, close-up photographs of the jovian moons and studies of the invisible plasma environment influenced by the magnetic fields of sun and earth. One of the major achievements of this decade is the capability of global monitoring and assessment of the long term variation of the earth's ozone layer which protects all forms of life on earth. Revolutionary concepts of long distance communication and broadcasting, remote sensing of the earth's resources and the real time monitoring of weather provide evidence of down-to-earth applications of space technology which no developing country can afford to ignore. The uniqueness of space technology is that it deals not only with the excitement of finding new answers in the realm of solar system, astronomy and cosmology, but also promises to be a very crucial tool for social and economic progress to improve the quality of life on earth.

The dawn of the first human civilization about 6000 years ago needed its directions from space and time. People cultivating the valley of Nile made use of the fact that the flood of the river fertilized the soil annually. They used the first appearance of the bright star Sothis in the high summer as an indicator of the new agricultural period when the flood in the Nile could be expected. This helped in continuing agriculture at the same place for thousands of years. It is during this period that the need to span the time scale from days to years led to the evolution of complicated calendars and sophisticated systems of numbers. Similarly the first sailors of Antiquity were guided by the Polar Star which indicates the direction North and its elevations above the horizon express the geographical latitude. Hence astronomy, necessary for survival and progress, became the mother of all sciences establishing the intellectual foundation of all great civilizations.

THE RELEVANCE

At the very fundamental level, the learning of space science fits the perpetual human urge for new and exciting knowledge as a part of the cultural process of enrichment of the human mind. Spectacular space events, like the landing of man on moon which represent the culmination of a complex endeavor of the highest intellectual order, have served to kindle public interest and enthusiasm in a way that no other scientific achievement has done so far. On the other hand one also witnesses elements of an age of ignorance and superstition related to cosmic phenomena such as lunar and solar eclipses, auroral displays, cometary apparitions, meteors, etc. even among the so called 'enlightened' people in spite of splendid strides made in space exploration. True enlightenment is essential through properly oriented education incorporating elements of modern knowledge accruing out of space exploration. As an illustration, in India through orchestrated programmes using media coverage, timely information dissemination about the total solar eclipse of October 24, 1995, its scientific basis and different facets of its spectacular visual effects led to a better awareness in the public compared to that for the previous total solar eclipse of February 16, 1980 and did much to drop the old beliefs and myths about such eclipses.

The development and operation of various types of space platforms have revolutionized applications in the fields of weather forecasting, global monitoring of the earth's resources, communication and broadcasting. The innovative technology and sophisticated sensors, probes and other developments have paved the way for scientific investigation <u>of</u> space, <u>from</u> space and <u>in</u> space. From the phenomenon of the earth's environment to evolution of the universe, space technology provides a novel avenue for an integrating scenario hitherto unknown in the history of science and technology.

While studies of space include topics related to solar and space physics, planetary environments, cosmology etc., research from space platforms involve the use of remote sensing techniques from space to study the earth sciences, the solar system and astrophysical phenomena. Studies in space provide a totally new avenue to scientific research related to the unique microgravity environment of spacecraft. The study of the behaviour of objects in space compared to their behaviour on earth became possible only with the advent of space technology. The microgravity condition helps research in fundamental science including human physiology and fundamental biology. The new physics which can be studied in space stems from the fact that the microgravity situation eliminates gravity driven convection, sedimentation, buoyancy, hydrostatic pressure, etc. Also the exposure of living organisms to the space radiation forms fundamental topics in radiation and exo-biology. Apart from these scientific applications, microgravity of space also holds great promise for the future making it possible to develop pure crystals, high strength alloys, pharmaceuticals, and other special materials. The fundamental studies of the space environment, i.e., electromagnetic and particle radiation, conditions of vacuum and microgravity led to the development of modern space systems for commercial use in communications, earth's resources survey, meteorological and environmental phenomena which have improved the quality of life on earth considerably.

Apart from the direct use of space science and technology for the above applications, there are a number of spin-offs to various industries in developing new technologies for improved standard of products. Industries not only participate in space endeavours by developing challenging technological subsystems but also gear up to utilize the new materials manufactured in space. For example, the development of life support systems for astronauts led to new techniques in food processing and environmental control and protection. The improved self lubricating bearings developed for space with less frictional losses, wear and tear have been capitalized upon by industries for use in the aeronautical and automobile sectors.

THE CHALLENGES

Of all the countries in the world, more than 120 are categorized as developing. Out of the total population of 5.3 billion in 1990, about 3.9 billion (73%) live in developing countries. The literacy in developing countries varies from about 50% in Africa to about 85% in South Africa. Scientists and engineers engaged in R&D activities in developing countries constitute about 200 per million as compared to the figure of 4000 per million in developed countries. The ratio of R&D expenditure to GNP is about 0.5% and 3% respectively, for developing and developed countries. By the year 2000, there will be 2.5 billion persons in the age group 5-24 years out of the predicted total of about 6 billion. While the present expenditure on education varies from 0.5-12% of GNP, UNESCO has recommended an expenditure of at least 4% of GNP for all countries (Badran, 1993). Within this scenario, a definitive fraction going from universal education to science education and then to space education needs to be ensured. Another main problem is the availability of adequate and trained manpower for space science education; the student-teacher ratio in many developing countries is about 1 in 70 as compared to about 1 in 20 in developed countries.

When it comes to investment, several questions crop up, primarily stemming from a lack of appreciation of the relevance by decision makers and their pre-conceived notions, particularly in developing countries. Such notions may arise from, for example, the confusion between astrology and astronomy, the view of space being a high technology endeavour of a few selected superpowers, or of space technology being useful only in warfare and not in any down-to-earth applications, and the view point that it is a subject which does not have any job potential or prospects in career development.

Space Science teaching poses many problems. For acquiring a clear concept of the subject advanced audio visual aids and simulation software need to be used, as conventional methods have serious limitations. Accordingly high levels of investment become necessary for imparting space education at various academic levels. The highly dynamic nature of the syllabus needs frequent updating of the course material as well as training of the trainers.

THE OBJECTIVES

Against the above background, what goals can one set for space science education? First, of course is the conveying of the vast body of knowledge of the space related phenomena to a larger cross-section of the public through methods which are easily understood and assimilated. Secondly, there is a need to maintain a suitable body of scientific and technological skill within society in the broader context of human resource development. Education brings awareness and develops public competence, resulting in a better appreciation and review of investment in the highly technological area of space research programmes.

Some of these objectives could be suitably oriented to suit local situations. NASA's initiative to offer space experiments to be undertaken by school children has been a bold and definite step in space education. NASA has a 2-year programme to design, develop, build and launch payloads weighing about 140 kg as part of student explorer demonstration initiative which is managed by the 75 member universities space research association. Other similar examples include (i) the selection of 22 experiments submitted to ESA by university students to conduct 20 sec. microgravity experiment in aircraft, (ii) the Moscow Aviation Institute and CNES, France organisation of space camps where students design small rockets to study flight dynamics, propulsion and management, (iii) Australian Space Research Institute (ASRI) run by students launching 6m long sounding rockets powered by liquid oxygen-kerosene engine, (iv) the science day (February 28) celebrations at ISRO Satellite Centre (ISAC), India, organizing quiz competitions on space themes, exhibitions, on-the-spot painting competitions, etc., (UN, 1994,1995).

THE MEANS

For complex multi-disciplinary subjects like astronomy, space science and cosmology, the education material, particularly for public awareness and for primary school children should be simple and illustrative such that it builds interest in the young minds. Several methods of imparting space education have been stressed which include interactive audio-visual shows, video sequencing of events such as MIR repair, space walk, etc., simulation programmes, computer aided video creating a virtual laboratory where students interactively conduct experiments and can make measurements and do calculations etc. Interests of the public and children are easily aroused through science fiction books, a few of which written by renowned authors such as A.C. Clarke, F.Hoyle, J. Verne, I. Asimov serve to generate curiosity about space in children and the younger generation. Amateur clubs, professional organizations and planetaria organise science films, video shows, training camps and workshops as part of popularization and promotional activities. In India summer workshops are arranged for Amateur astronomers making small telescopes. In addition to these methods, space technology can itself be used for effectively imparting space education. One of the techniques used in India is the INSAT satellite transmitting instructional courses with a talk-back facility for interactive learning (Rao, 1995).

A number of developed countries have many facilities for space science education. For developing countries, international cooperation may help in strengthening the activities of space science education and training programmes being pursued at identified centres. Such cooperation would help in developing advanced teaching aids, sharing of experience in teaching, curricula development and training of teachers (Jasentuliyana, 1995).

CONCLUSION

The organization of space science education activities is as complex and challenging as undertaking a space mission. There are many approaches in the global scenario from which it is difficult to select the most appropriate one for a developing country. Highly innovative and pragmatic ideas are required for introducing space education, keeping in mind the overall goals of pursuing space research for the benefit of the common man. The adoption of space science education programme with local talents, facilities and resources seems to be a major challenge. Close cooperation would be needed between national space agencies, research institutions and universities for the development of space science education programmes in any country. Also the establishment of linkages with international agencies, academic bodies, etc., would help both in initiating and sustaining such a programme.

REFERENCES

Badran, A., Basic and applied sciences in the world – past strategies and present status – the role of UNESCO, in *Strengthening basic and applied science education in developing countries*, Proceedings of an international seminar in Trieste, Italy, April 5, 1993 edited by R. R. Daniel and V. Ravichandran, pp 1-49, VINCAAR Publishing Co. , Madras, India (1993)

Jasentuliyana, N., Basic space science and developing countries, *Space Policy*, **11(2)**, 89(1995)

Rao, U. R., Space and education, in *Space for education, healthcare and environment monitoring*, Proceedings of Special Plenary Session, International Astronautical Federation, 46th Congress, Oslo, Norway , 1995 (1995)

UN Report: *Highlights in Space*, Progress in space science, technology and applications, international cooperation and space law, pp 49-51 (1994), ibid. pp 48-50 (1995)

Pergamon

Adv. Space Res. Vol. 20, No. 7, pp. 1335–1339, 1997

Printed in Great Britain
0273-1177/97 $17.00 + 0.00

PII: S0273–1177(97)00727–8

SPACE SCIENCE EDUCATION—AN INTERDISCIPLINARY AND INTERNATIONAL PROGRAMME

M. J. Rycroft

International Space University, Parc d'Innovation, Boulevard Gonthier d'Andernach, 67400 Illkirch, France

ABSTRACT

The ISU's Master of Space Studies (MSS) programme is a new, unique and interdisciplinary postgraduate course of eleven months duration. It is designed for those who will lead and work in the international space programmes of the next century. It considers not only many aspects of the space sciences and their applications, but also most engineering, business, management, policy and law issues relevant to the activities of both space agencies and space industries. In the first term, fundamental issues are covered, and MSS participants from many countries around the world select some topics of their choice; within the sciences, attention is focused on basic physics, the space environment and life sciences. In the second and third terms, advanced subjects ranging from astronomy to planetary sciences, microgravity sciences, telecommunications and remote sensing of the Earth's surface and atmosphere are available. For two months in the second term, each participant carries out a project at an affiliate campus or an associated institution, and prepares a written report on this. Throughout the year, all participants play an active part in a team design project on a chosen theme; a substantial written report is prepared, and the results are also presented orally. Thus the aim of the MSS programme is to educate space generalists, yet generalists who are knowledgeable in all the relevant disciplines, including science.

INTRODUCTION

An exciting event on the international stage of space education was the commencement in Strasbourg of the ISU's one year Master of Space Studies (MSS) course in September 1995. This followed on from the successful annual – and peripatetic – ISU Summer Session Program (SSP). Of ten weeks' duration that programme started in 1988. Three key words of all International Space University (ISU) activities are international, intercultural and interdisciplinary.

The ISU is an innovative educational institution dedicated to excellence in space education in a changing world. This is especially necessary following the end of the Cold War era, with the concomitant change from "technology push" to "user pull". The many diverse requirements of the space user community are now driving technological developments in the space industries. New opportunities (supported by the users of space data) are overtaking innovative possibilities (determined by the planners and providers) on the highway to the third millennium, the space millennium.

Further the ISU is developing a Professional Development Program (short courses and seminars), activities as an independent neutral forum (conferences, workshops and discussion groups), and research.

MASTER OF SPACE STUDIES (MSS) COURSE

The MSS program at the ISU is a unique postgraduate course both for young graduates and for more mature space professionals. It treats all major space-related disciples – both technical and non-technical – in an integrated way. Thus it offers a broad, challenging and mind-expanding degree course. With a global perspective, it emphasises the interactions between the disciplines.

The MSS course is designed to educate the staff who will play key roles – of influence and responsibility – as employees of the space industries, of national and international organisations, and in academia.

In September 1995 thirty very individualistic students having very different educational backgrounds, from fourteen countries, joined this course which is carried out entirely in English. The academic year is planned in three terms – term 1 (16 weeks, September to December), term 2 (12 weeks, January to March) and term 3 (16 weeks, April to July). There are three major components to the course, as shown in Figure 1. Lectures (term 1 and first half of term 3), the Individual Project (term 2) and the Team Design Project (throughout the year, yet reaching a crescendo of activity in the second half of term 3).

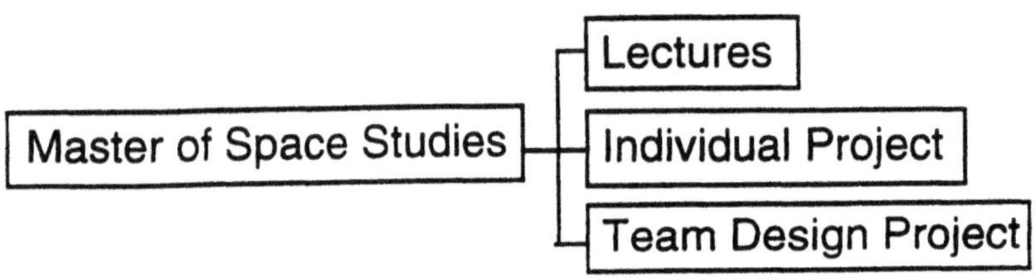

Fig. 1. Structure of Master of Space Studies course.

MSS CURRICULUM

There are fourteen interrelated components of the MSS curriculum. In Figure 2, these are organised as items in a feast. During the academic year, the students have the set menu of seven items and they also choose within all seven of the à la carte items.

The pedagogical philosophy and purpose of the most significant of these items is outlined here.

Preparatory classes, introductory lectures: to teach a basic knowledge of fundamental concepts to all students, whatever their own discipline.

Seminars, selected topics, advanced subjects and contemporary and future issues (mainly in term 3): modular courses, typically of 5 to 10 lectures, to give a deeper insight into more specialised subject areas, and to learn interactively.

Foreign languages: weekly classes to improve English and/or French.

Individual Projects, professional placements: to visit an ISU Affiliate campus (of which there are twenty five around the world) or another host institution (for example, a government research centre or a space company) in order to learn by experience; a written report is prepared and an oral presentation given.

Team (Group) Design Project: to put into practice what has been learnt, to develop an innovative conceptual study of value to the wide space community, analysing many issues from different perspectives, to enhance teamworking skills, group problem-solving and decision making skills, and to create a global network of professional relationships. (Details have been given by Rycroft et al., 1996.)

ISU Symposium: this major annual event focuses on a theme which includes the topic of the Team Design Project for the year.

Within this carefully integrated MSS course, the following disciplines may be studied:

In the field of management and social sciences
- economics, business and management of space programmes,
- law and policy of space activities,
- space and society, including anthropology, education, ethics, history, literature, philosophy and sociology.

In the field of engineering, system and technologies
- mission analysis and systems architecture,
- launch vehicles and transportation systems,
- orbital mechanic,
- command, control and communications systems,
- space architecture and bases.

In the field of sciences and applications
- space physical sciences, including astronomy, the planets and solar-terrestrial physics,
- space life sciences, both cellular and human,
- space resources and processing under microgravity conditions,
- Earth observation, including remote sensing and Geographic Information Systems,
- Earth-oriented space applications, including telecommunications, global positioning systems and global navigation systems.

Each student works with an Academic Advisor throughout the year. The Advisor's role ties together many aspects of the programme, as illustrated in Figure 4 (from an idea by Susan Berry).

Fundamental issues: a core course, and a comprehensive course (both in term 1), to give a balanced – and thorough – understanding in all space-related fields (the space environment, the space area, space missions, the Internet), both technical and non-technical (see Figure 3), and to develop cross-cultural communication skills.

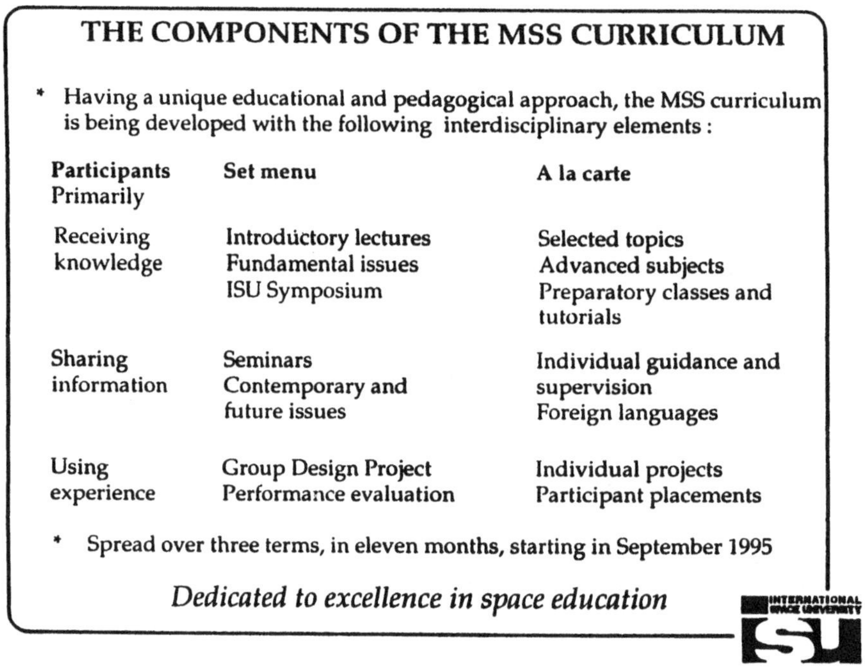

Fig. 2. Components of the MSS curriculum.

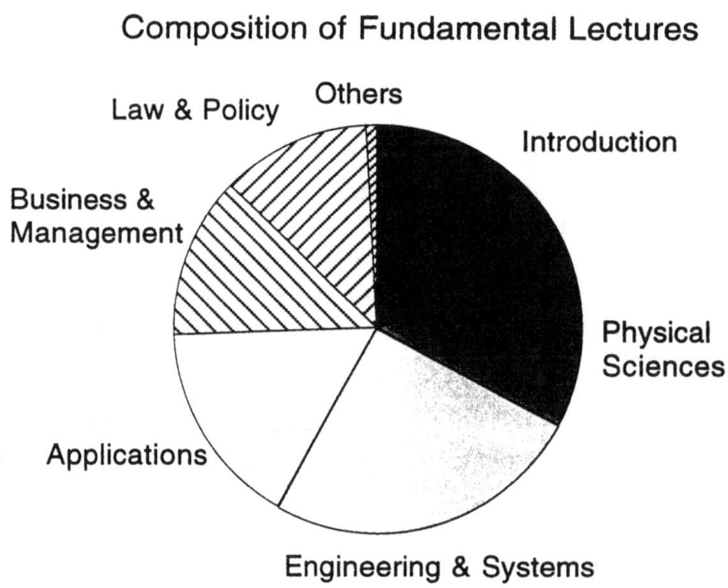

Fig. 3. Distribution of Fundamental Lectures in the 1995/96 MSS programme.

Seminars, selected topics, advanced subjects and contemporary and future issues (mainly in term 3): modular courses, typically of 5 to 10 lectures, to give a deeper insight into more specialised subject areas, and to learn interactively.

Foreign languages: weekly classes to improve English and/or French.

Individual Projects, professional placements: to visit an ISU Affiliate campus (of which there are twenty five around the world) or another host institution (for example, a government research centre or a space company) in order to learn by experience; a written report is prepared and an oral presentation given.

Team (Group) Design Project: to put into practice what has been learnt, to develop an innovative conceptual study of value to the wide space community, analysing many issues from different perspectives, to enhance teamworking skills, group problem-solving and decision making skills, and to create a global network of professional relationships. (Details have been given by Rycroft et al., 1996.)

ISU Symposium: this major annual event focuses on a theme which includes the topic of the Team Design Project for the year.

Within this carefully integrated MSS course, the following disciplines may be studied:

In the field of management and social sciences
- economics, business and management of space programmes,
- law and policy of space activities,
- space and society, including anthropology, education, ethics, history, literature, philosophy and sociology.

In the field of engineering, system and technologies
- mission analysis and systems architecture,
- launch vehicles and transportation systems,
- orbital mechanic,
- command, control and communications systems,
- space architecture and bases.

In the field of sciences and applications
- space physical sciences, including astronomy, the planets and solar-terrestrial physics,
- space life sciences, both cellular and human,
- space resources and processing under microgravity conditions,
- Earth observation, including remote sensing and Geographic Information Systems,
- Earth-oriented space applications, including telecommunications, global positioning systems and global navigation systems.

Each student works with an Academic Advisor throughout the year. The Advisor's role ties together many aspects of the programme, as illustrated in Figure 4 (from an idea by Susan Berry).

Fundamental issues: a core course, and a comprehensive course (both in term 1), to give a balanced – and thorough – understanding in all space-related fields (the space environment, the space area, space missions, the Internet), both technical and non-technical (see Figure 3), and to develop cross-cultural communication skills.

THE COMPONENTS OF THE MSS CURRICULUM

* Having a unique educational and pedagogical approach, the MSS curriculum is being developed with the following interdisciplinary elements :

Participants Primarily	**Set menu**	**A la carte**
Receiving knowledge	Introductory lectures Fundamental issues ISU Symposium	Selected topics Advanced subjects Preparatory classes and tutorials
Sharing information	Seminars Contemporary and future issues	Individual guidance and supervision Foreign languages
Using experience	Group Design Project Performance evaluation	Individual projects Participant placements

* Spread over three terms, in eleven months, starting in September 1995

Dedicated to excellence in space education

INTERNATIONAL SPACE UNIVERSITY ISU

Fig. 2. Components of the MSS curriculum.

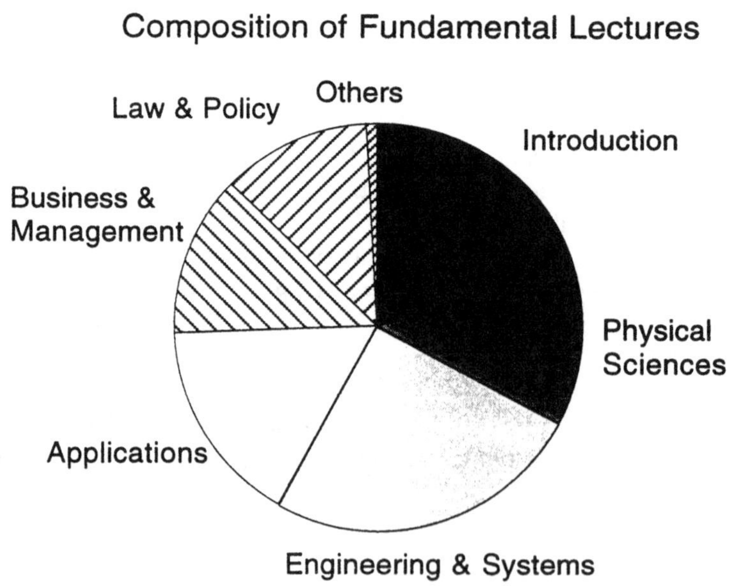

Fig. 3. Distribution of Fundamental Lectures in the 1995/96 MSS programme.

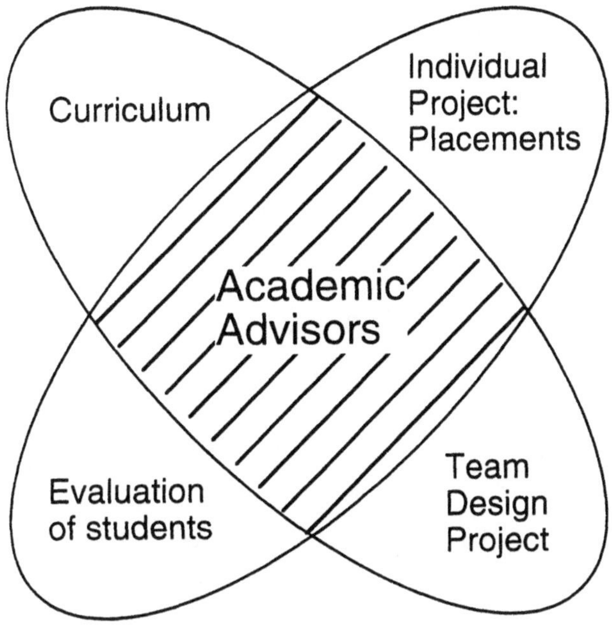

Fig. 4. Diagram illustrating the central role of Academic Advisors in the MSS programme.

Some detailed changes to the curriculum will be made in successive years, based upon the experience of the previous academic year.

CONCLUSION

The new Master of Space Studies (MSS) course of the ISU has been outlined. Through it, students:
- gain a thorough knowledge in all major space-related disciplines and their inter-relationships,
- become part of an active worldwide network of professional colleagues,
- carry out an Individual Project during a professional placement period, enhancing self-learning and research skills,
- pursue one, or more, area(s) of specialisation in space activities,
- gain state-of-the-art information,
- participate in a Team Design Project, confronting the real challenges of creating a conceptual space project, and
- develop the personal qualities and professional skills needed to communicate and to work effectively with colleagues in other countries.

Only time will tell how valuable the MSS course is. That will be in ten to twenty years from now when MSS graduates will work in key positions in many areas of space activities around the world.

ACKNOWLEDGMENTS

The author is especially grateful to Professors Michel Bousquet and Ram Jakhu for many hours of collaboration in the detailed planning and implementation of the first MSS programme.

REFERENCES

Rycroft, M., M. Cho, R. Thomas, and S. Berry, Post-graduate interdisciplinary education for cooperative international space activities, 96-m-27, 20th. Gifu ISTS and 11th. IAS, Japan, pp.5 (1996).

Pergamon

Adv. Space Res. Vol. 20, No. 7, pp. 1341–1349, 1997

Printed in Great Britain
0273-1177/97 $17.00 + 0.00

PII: S0273-1177(97)00728-X

SPACE EDUCATION

Adigun Ade Abiodun

Office for Outer Space Affairs, United Nations, Vienna, Austria

ABSTRACT

An essential pre-requisite to a successful space technology applications programme is the building of various indigenous capacities, particularly human resources. Efforts to accomplish such a capacity-building must be devoted, at the local level, to the development of necessary high-level knowledge and expertise in space science and technology fields. Such a programme must also focus on long-term in-depth education and research opportunities in the developing countries, where the beneficiaries would gain an in-depth understanding and appreciation of not only the application potentials of a given technology but also an insight into why and how the technology works the way it does. In recognition of such a pre-requisite, it is universally acknowledged that if effective assimilation of space science and appropriate application of space technology are to succeed in the developing countries, and particularly if such a discipline as satellite remote sensing is to transcend its current image of being a technology-driven tool into a user-driven one, efforts must be devoted, at the local level, to the development of necessary high-level knowledge and expertise in requisite space science and technology fields. The justification for such an in-depth education is not far-fetched particularly as one reflects on the myriad of space science and technology activities that are both in progress and are planned. Aspects of these are reflected in this paper.

UNDERSTANDING OUR ENVIRONMENT

The observation of the Earth and its immediate environment from the vantage point of outer space has become routine. Today, Earth observation satellites provide copious data that are processed, analysed and subsequently used by many countries to decide on how to address a variety of issues that affect their natural resources and their environment. Similarly, meteorological satellites and their associated sensors offer humankind the tools regularly to monitor the state of the Earth's atmosphere, including its pressure, temperature, humidity, wind velocity and chemical constituents. As an integral part of the planetary life-support system, the atmosphere has an overwhelming impact on most human activities including energy development, human habitat (development of cities and design of buildings), water resources development, human health, public safety, transportation, construction activities and recreation. In the area of public safety, for example, hurricanes and severe storms can now be located, their directions and intensities determined, and individuals warned in advance of impending disasters.

Observations from space have also demonstrated the versatility of the atmosphere as a transport and distribution medium, such as the global distribution of the ashes and dust that resulted from the volcanic eruptions of Mt. Pinatubo (the Philippines) in 1991. Such observations have also improved human understanding of the global chemical cycles as well as our knowledge of the impact of local/regional activities, including those attributable to natural and anthropogenic emissions. The intervention of humankind in the natural self-regulating processes within the atmosphere is creating a major disequilibrium in this fragile medium and is producing climatic changes which may threaten the continuing habitability of planet Earth. Thus, for our own safety on planet Earth, it is imperative that we have a full and complete understanding of the state of the atmosphere at all times and in all localities. Adequate knowledge and long-term research exposure by scientists from the developing countries are essential in order that this segment of the global community could contribute effectively to the solution of this and other global events.

As we approach the 21st century, the competition is tightening and major technological foundations that will take human-kind into the future, are being laid. The challenge for those that have begun the race as well as those that are still on the sideline, particularly the developing countries, is how to secure a technological and an economical competitive position, and thus *escape domination in any form in the years ahead.* For example, the race is on to build a future world that is wireless; to achieve this goal, three private entities in Europe, Japan and USA are planning to place a total of 994 communication satellites in the low-Earth orbit before the year 2000. These developments will enable customers to receive calls anywhere in the world except near the North and South poles. Electronic mail from laptop computers and personal communication devices will also become a reality. Similarly, geo-positioning, the technology of knowing where you are, at any time, on Earth or space, is now a reality - made possible by USA's Global Positioning System (GPS) of satellites with a constellation of 24 satellites in low Earth orbit (20,000 km altitude), and by Russia's constellation of 12 satellites (24 planned in 1996) within its Global Navigation Satellite System (GLONASS).

Certainly, future space systems will vary in shape, size and mission. More nations and entities will own, launch and operate their own satellites for a variety of purposes; and there will be a wide array of launch vehicles to choose from. Today, the list of space-capable countries continues to grow with many more achieving launch capabilities. International co-operation in space is being manifested through several collaborative arrangements. The most prominent of these is the joint effort by Canada, European Space Agency, Japan, Russia and USA, together or in groups, to build the international Space Station, the platform that would make space industrialisation, within the microgravity environment of outer space, a reality, in the first quarter of the 21st century. It is now believed that high value materials such as precision latex micro-spheres, electronic materials, various kinds of pharmaceutical products, optical fibres, and highly specialised alloys can be produced, in their purest forms, in the micro-gravity environment of an orbiting spacecraft.

Space industrialisation also holds promise for the production of new antibiotics, antibodies and other treatments for cancer, malaria and other diseases including the growth of collagen fibres to be used in the repair and replacement of human connective tissues. Of major interest is the performance of various lubricants and the growth of crystals in outer space for semi-conductor devices. These possibilities would certainly improve the efficiency of the production process as well as the quality of goods produced on Earth. The attendant economic impact could be very significant and far-reaching.

Human venture into outer space has also introduced new dimensions into many Earth-based processes and procedures. The offshoots of these developments is a host of invaluable products such as the telemetry system in the heart pace-maker, cordless mechanical drilling and cutting tools, and photo-voltaic cells. In addition, the limited space on board a space vehicle requires that all space-bound instruments and tools be as light and compact as possible without any loss of efficiency. The spin-off benefits of developing, designing and manufacturing successfully such space-bound equipment include miniaturised, inexpensive and versatile consumer products - such as television sets, radios, cameras, video-games, microcassettes and portable receive-only TV antennas.

Furthermore, the close working relationships on space programmes between governments, universities and private and public institutions, particularly in space-faring nations, have enhanced the scientific and technical capacities and capabilities of institutions of higher learning in these countries, and have expanded their contributory roles in the social and economic development of their societies. The on-going technology developments and exploitation processes in these societies suggest that future economic systems here on Earth would be based on a redistribution of both Earth-bound and space-based resources and other benefits associated with space exploration and utilisation.

The aforementioned achievements became a reality through the systematic infusion of time and financial resources into the development of necessary skills and knowledge which were subsequently judiciously utilised. How will the developing countries fit into this emerging future?

DEVELOPING COUNTRIES TODAY

A number of developing countries, such as Brazil, China and India, are integrating research efforts into their overall national development programmes. By their very nature, successful research efforts are often conceived, not in a narrow scope, but in a wider concept in order that they can influence the different aspects of a nation's social and economic plans. In order to be effective, such research programmes should be designed in stages; each succeeding stage should be able to demonstrate advances in the research process as well as a corresponding higher level of attainment/ achievement. Each stage should not be terminal, but should be a precursor of future trends. Brazil, China and India have adopted some of these approaches and have invested their manpower and financial resources in their respective national space research programmes. Subsequent achievements of these countries in electronics, satellite development and propulsion technology have led to their being collectively referred to today by the industrialised community as "the emerging industrialised countries".

A sound science and technology culture is a major prerequisite for a successful technology development effort. To this end, Malaysia has embarked on developing the public's broad interests in science in general, including space science and technology and its related applications, through its recent establishment of a planetarium and an astronomical observatory (Feldberg, 1994). This facility should provide an opportunity for interested individuals within the society to educate themselves about their relationship with the rest of the universe, promote general awareness, and serve as stimulus for general interests in physical sciences, including space science and technology. Similar programmes have been established in Indonesia, Thailand and several countries in Latin America. In addition, a number of developing countries have also attained the capabilities for designing and building necessary ground-based communications antennae and associated facilities. A few others, such as Chile, Mexico, Republic of Korea and South Africa are contributing to research and development of small scientific satellites and their payloads as a means of acquiring knowledge and skill in space research and development.

Today, many developing countries, particularly in Africa, are still saddled with a variety of local communication regulations, and the monopolies enjoyed by their inefficient PTTs' monopolies for both internal and external communications services. As an alternative, many research institutes and universities in Africa are working, through their counterparts in Europe and North America, to be part of the information age not only for the exchange of scientific ideas but also for the delivery of medical advice and services.

The above conditions notwithstanding, many developing countries are experiencing major difficulties in forging ahead in various aspects of space technology related areas and have had to resort to a combination of short term training courses, on-the-job training, either locally or abroad as well as a dependence on foreign experts for the definition, design and execution of most space applications projects. In most cases, such projects cannot be sustained by local counterparts whose knowledge of the fundamentals is very limited.

Additional contributing factors to the above state of affairs are the following:

- the limited funding available from governments
- the shortage of qualified human resources at the local level
- the shortage of education/training facilities
- lack of information about latest technological developments in space-related applications, products, services and equipment
- limited access to basic facilities (such as those needed for communication, information exchange, computing)
- lack of adequate information for formulating scientific and technological policy and for effecting an efficient implementation of space-related projects.

Formulation, updating and efficient implementation of space-related projects require professional personnel with appropriate skills and a fundamental scientific knowledge and understanding of the principles involved. Such personnel should be able to adapt the technology to match the local conditions and working environment in order to obtain maximum results.

Today, the international community fully recognises that an essential pre-requisite to a successful space technology applications, especially in the developing countries, is the building of various pertinent indigenous capacities, particularly human resources, within each country/region (Prince *et al*, 1990) and (World Bank, 1991). Visionary leaders in the developing countries are equally conscious of this need.

MUTUALLY BENEFICIAL PROGRAMMES

In particular, the developing countries need to develop their own capacities and capabilities in space science and technology. Accomplishing the latter will enable these countries to (i) share in the direct and indirect benefits of space exploration, and (ii) equip themselves to contribute effectively to and participate in national, regional and international programmes, particularly in communications systems, climate research, ocean studies, and land-use and forest inventory. Prominent among these programmes are (a) the internationally co-ordinated space-based Mission to Planet Earth which will chart pollution, deforestation, the greenhouse effect, ozone depletion and threats to the Earth's environment, (b) Agenda 21: a variety of programmes (adopted in Rio at the 1992 UNCED Conference) that could lay the foundation for a global partnership in order to ensure, through sustainable development, the future of our planet, (c) the World Weather Watch and the World Climate Programme, and (d) the International Geosphere-Biosphere Programme which strives to describe and understand the interactive physical, chemical and biological processes that regulate the Earth system.

For example, the successful operation of an international remote sensing system (Abiodun, 1993) or an International Global Observing System (IGOS) (CEOS, 1996) which is being proposed by the Committee on Earth Observation Satellite (CEOS) presupposes that it would be mutually beneficial for both the industrialised and the developing countries. For the latter, such a step will be an important one, since they will become active partners in a global venture, such as on environmental monitoring, through the modern techniques of space systems. In addition, the related activities could provide impetus for the creation of new institutions and systems or the expansion of existing ones and subsequently facilitate the handling and use of space acquired data as part of an overall long-term development strategy in these countries.

To accomplish the above, a higher level of knowledge and expertise than exists today is needed by educators and research and application scientists in the developing countries so that they can make significant contributions to the solution of global, regional and national environmental problems and become active partners in these global ventures. In particular, educators and scientists in the developing countries should participate in the definition of remote sensing instruments, and in the investigation of the weaknesses and difficulties in current and future instrument designs including the definition of techniques for their improvement. They should contribute to the development of calibration techniques for Earth-observation data including the development of software for image assessment, the establishment of assessment criteria and the definition of calibration strategies with the aid of calibration targets and targets of opportunity. They should also contribute to the evaluation of Earth-observation data and to the assessment of their value in various Earth science disciplines. These capabilities can only be acquired through long-term intensive education.

The need for such intensive research and education programmes has been receiving increasing attention at the United Nations, and has resulted in a number of initiatives - four of these include (a) the United Nations/Sweden International Course series on Remote Sensing Education for Educators, (b) the United Nations/European Space Agency workshops on Basic Space Science, (c) long-term fellowships for in-depth training, and (d) the establishment of Centres for Space Science and Technology Education (UN document, 1995).

Remote sensing education for educators

In response to the recommendation of UNISPACE 82 and those of the United Nations International Meeting of Experts on the Development of Remote Sensing Skills and Knowledge, held at Dundee, Scotland, in 1989, the Government of Sweden in 1990, through the Swedish Board for Co-operation Agency (SIDA), initiated jointly with the Office for Outer Space Affairs the annual six week course series on Remote Sensing Education for Educators. The course is hosted by the Department of Physical Geography, Stockholm University, and the Swedish Space Corporation (SSC Satellitbild) at Kiruna and co-organised with the Office for Outer Space Affairs, through the United Nations Space Applications Programme. The course is directed at educators from African, Asia-Pacific, Latin America and the Caribbean and Middle East universities.

The main objective of the course is to develop the practical knowledge and skills of educators from developing countries in remote sensing technology and to equip them with the skills needed to introduce the discipline, as appropriate, into the educational curricula of universities and institutes in their home countries. The participants are taught project planning and management and are exposed to the various aspects of satellite data analysis and applications, using data and materials from their own countries, with particular emphasis on: (a) remote sensing and global change; (b) remote sensing applications: limitations and potential in developing countries; (c) remote sensing training strategy in developing countries; (d) remote sensing satellite systems; (e) visual and digital interpretation of satellite images; and (f) land-use planning and environmental monitoring, including vegetation mapping, change monitoring and water resource development.

Participants are also provided additional opportunities for practical training in the visual interpretation of remote sensing data, through numerous practical exercises and field trips to the Abisco Scientific Research Station (100 km north of Kiruna). The course programme is also supplemented with visits to SSC Satellitbild facilities and laboratories at Kiruna, the European Space Agency (ESA) Salmijarvi Ground Station and to the Esrange satellite station of ESA.

Basic space science workshops

In 1991, the United Nations, through its Space Applications Programme, and the European Space Agency initiated an annual workshop series on Basic Space Science for the developing countries; it has been held in India (1991), Costa Rica (1992), Nigeria (1993), Egypt (1994) and Sri Lanka (January,1996). The concluding workshop in this series will be organised in Germany in September 1996.

These workshops have focused, in particular, on education, research and scientific co-operation as pre-requisites to the development of space science and technology capabilities at the national, regional and international levels. In this context, the workshop programmes have addressed space science education and policy, solar-terrestrial interactions, planetary science, ground- and space-based observatories, space astronomy, astrophysics and cosmology.

A number of projects which are enhancing regional and international co-operation have resulted from the deliberations of these workshops; these include (i) the inauguration in Sri Lanka, in January 1996, of an astronomical telescope donated by Japan; (ii) the operation of an astronomical observatory in Honduras; (iii) the on-going development of an Inter-African Astronomical Observatory and Science Park on the Gamsberg in Namibia; and (iv) the on-going upgrading of the Kottamia Observatory in Egypt.

Long-term fellowships for in-depth training

In furtherance of the recommendations of UNISPACE 82, and in order to enhance both the research and practical experience of scientists and engineers in the developing countries, the United Nations instituted, through its Space Applications Programme, the annual administration of long-term fellowships for in-depth training. These fellowships are currently tenable at the European Space Research Institute and Technology Centres of the European Space Agency and at Brazil's Institut Nacional de Pesquisas Espaciais (INPE).

The ten (10) fellowships offered by Brazil are each tenable for a period of seven months and are for research and applications in remote sensing technology. The five ESA fellowships are each tenable for a one calendar year of research and study in (i) space antennas and propagation, (ii) communication systems, (iii) remote sensing information systems (2 fellowships), and (iv) remote sensing instrumentation. These fellowship awards cover monthly allowances for room and board, books, local travel and health benefits which are provided by the host institutions. Two hundred and seventy five (275) participants have benefited from this fellowship programme since its inception in 1983.

Centres for space science and technology education

As the need to develop indigenous capacity became paramount, the General Assembly (GA), in its resolution 45/72 of 11 December 1990, endorsed the recommendation of its Committee on the Peaceful Uses of Outer Space (COPUOS) that: ***the United Nations should lead, with the active support of its specialised agencies and other international organisations, an international effort to establish regional centres for space science and technology education in existing institutions in the developing countries.***

The host countries for the Centres are: India (Asia and the Pacific), Brazil and Mexico (Latin America and the Caribbean), Morocco-Francophone countries and Nigeria-Anglophone countries (Africa). Eight central, eastern and south-eastern European countries plan to establish a similar centre that will operate in a network mode. There is a plan to establish a similar Centre in the Middle East. The India Centre began its first programme on remote sensing and GIS on 1 April 1996 and will begin the communication and satellite meteorology programmes in 1997. The Centre in Brazil and Mexico should open before the end of 1996. The Africa Centre should begin operating by 1997.

Work Programme and Model Curricula: The activities at each Centre will be undertaken in two major phases. Phase 1 will emphasise the development and enhancement of the knowledge and skills of university educators and research and application scientists in both the physical and natural sciences as well as in analytical disciplines. These activities will be accomplished through rigorous theory, research, applications and field exercises over a nine-month period as laid out in the curricula of the education programme of each Centre. Phase 2 will focus on ensuring that the participants make use of the skills and knowledge gained in phase 1 in their pilot projects.

Each Centre will also foster continuing education programmes for its graduates, and awareness programmes for policy and decision makers and the general public in its region.

The model curricula for the Centres provide for a 2 to 3-month obligatory common curriculum for all participating scholars, and a 6-7 month individual curriculum in (a) Remote Sensing and Geographical Information System (GIS), (b) Satellite Meteorology, (c) Satellite Communications and Geo-positioning System (GPS), and (d) Space and Atmospheric Sciences. Thereafter, each participating scholar will carry out a 12-month project in his/her own country, where and when the knowledge gained at the Centre would need to be put into practical use.

In addition to providing opportunities for each scholar to gain necessary knowledge, research experience and application skills in his/her chosen area of space science and technology, each Centre's programme also requires these scholars to complete an obligatory common assignment at the beginning of their studies. The successful completion of this initial obligatory common assignment, which is the same for all participating scholars, is a pre-requisite to the enrolment of each scholar in his/her chosen field of study. This common module will provide all the scholars with an overview of the observation of the Earth and its environment from space and the use of data collected in such a process in atmospheric and terrestrial analysis. The obligatory programme will also expose the scholars to the physical principles of remote sensing; satellite orbital characteristics; operational sensors; satellite and ground based communications; the impact of global positioning satellites on the integration and construction of remote sensing and GIS databases and demonstration of selected environmental applications.

Each Centre shall aspire to be a highly reputable regional institution, which, as the needs arise, and as directed by its Governing Board, will grow into a network of speciality and internationally acclaimed affiliate nodes. The Centres and their nodes will earn this badge of honour through their contributions to the development of technologies that are appropriate to the solutions of the problems of their respective region, and to the advancement of knowledge in this ever expanding field. The model curricula for the Centres provide the bench mark of the academic and performance level necessary to maintain the international standard and character needed for the attainment of international recognition.

Data Management: An integral part of each of the Centres for Space Science and Technology Education is a Data Management Unit. Through such a Unit, each Centre will have direct linkages with existing relevant global data centres. Such linkages will enable the participating scholars to gain access to and utilise the data in the archives of a variety of data bases, particularly when undertaking projects/activities that could benefit from such an access.

An active participation and contribution of the developing countries in regional and global programmes require that they gain reliable and unimpeded access to global information networks. Today, in societies that have begun to invest in the future, many people base their hopes for education, profit and competitiveness in the ability of information technologies to provide the necessary electronic connections. The information age is demonstrating daily that our neighbourhoods and where we do business do not necessarily have to be limited to our immediate physical environment. Indeed, information is now becoming the most important global commodity through which one can secure, with the aid of satellite services, employment and new markets, and improve the health and education of any community including particularly the rural areas. A call for action by African professionals to improve the existing state of information exchange within Africa, was articulated at the United Nations Regional Conference on Space Technology for Sustainable Development in Africa, held in Dakar, Senegal in October 1993 (CEOS, 1996). This recommendation of this Conference has since been translated by the United Nations, under the auspices of its Space Applications Programme, into a project proposal called COPINE: A Co-operative Information Network Linking Scientists, Educators and Professionals in Africa, which focuses on the implementation of a satellite-based information system (COPINE, 1995)

It is anticipated that COPINE would be replicated in other regions of the world and could be a significant supporting tool for the Data Management Unit in each of the Centres for Space Science and Technology Education.

The Participating Scholars: A sound academic background, experience and aptitude of each applicant for engaging in the different activities of the Centre cannot be over-stressed. The richness of these attributes will have a positive impact on the applicant's performance at the Centre. Toward this end, each applicant (university educator, research or application scientist) should have obtained, from an internationally recognised university/institution, a minimum of master's degree, relevant to his/her chosen field of study, followed by a minimum of 5 years of relevant practical/working experience. An applicant with a Ph.D. degree, relevant to his/her chosen field of study, and from an internationally recognised university/institution should also have completed a minimum of 3 years of practical/working experience.

Of equal importance is the future of the participating scholars in their own countries on the completion of their studies at the Centres. It should be emphasised that the overall mission of the Centres is to assist participating countries to develop and enhance the knowledge and skills of their citizens in relevant aspects of space science and technology in order that such individuals can effectively contribute to national development programmes. In order to ensure that appropriate and rewarding employment opportunities exist for these returning scholars, the sponsoring governments/institutions are obliged to (a) sponsor development-oriented activities that will gainfully utilise the newly acquired knowledge and skills of the returning scholars, and (b) provide appropriate infrastructure, and undertake requisite preparations and plans for their career on a long-term basis. The sponsoring governments are also obliged to guarantee that a returning scholar would remain in such a position with commensurate and progressive remuneration and other entitlement for a minimum of 3-5 years.

The Governing Board: Because resolution 45/72 specifically limits the role of the United Nations *to lead international efforts to establish these Centres*, it is apparent that once any of the Centres is inaugurated, its Governing Board will assume all decision-making and policy formulation responsibilities for the Centre.

In the context of these Centres, what is a Governing Board, and why is it necessary? A Governing Board is the overall policy making body of each Centre; it oversees every and all aspects of the Centre. It consists of member States (within the region where the Centre is located), that have agreed, through their endorsement of the Centre's agreement, to the goals and objectives of the Centre, and are fully committed to work, in co-operation with one another, for the well-being of the Centre. A Governing Board, composed as described above, is necessary for each of these Centres because member States and their own citizens are more familiar with their own peculiar needs, aspirations, capabilities and resources and are better equipped to find solutions to local problems that may surface. No part of the United Nations system, including the regional economic commissions, is equipped to address such an array of issues, particularly within the framework of these Centres. And because such a Centre has evolved through the efforts of the United Nations, the latter, including the relevant regional economic commission, shall serve the Centre and its Governing Board in an advisory capacity.

CONCLUSION

One of the global needs today is an adequate opportunity for the men and women of each nation on this planet to experience the joy of successfully exercising their intellect and of unleashing their energies for the management of the environment and for the harnessing of the natural resources of their country in particular and of planet Earth as a whole. Such an opportunity should provide each nation with the ability to understand, process, analyse and judiciously utilise the copious information being provided daily and on a continuous basis, by an array of satellites. Bearing in mind that "the quality and character of a man's perceptions are determined in part by limitations imposed by or opportunities available in his environment", it is an inherent obligation of each country to provide the enabling environment that could ensure its effective participation in those space and science technology programmes that can enhance its social and economic development.

REFERENCES

Abiodun,A. A.,. An international remote sensing system: a possibility. ***Space Policy***, **9(3)**, pp 179-184, (1993)

CEOS: 1996, Report of the ad hoc CEOS Meeting on the Space Component of an Integrated Global Observing Strategy, Seattle, Washington, USA, 27-29 March 1996

COPINE: A Co-operative Information Network Linking Scientists, Educators and Professionals in Africa, a project proposal prepared under the auspices of the United Nations Space Applications Programme, Office for Outer Space Affairs, Vienna, February, 1995

Gregory H. Feldberg, Malaysia in *Mercury*, pp 6 (1994)

S.D. Prince, Justice, C. O., and Los, S.O., Remote Sensing of the Sahelian Environment: A review of the current status and future prospects, Report (No. ISBN 92 9081072 6) prepared for the Commission of the European Communities, Directorate General VIII (Ispra: Joint Research Centre), (19900

UN Document A/AC.105/625 (13 December,1995) Report of the United Nations Expert on Space Applications to the Committee on the Peaceful Uses of Outer Space, Vienna, Austria (1995)

World Bank, Program on environment information systems in Sub-Saharan Africa: Status of the Program. Report of the Third Meeting of the Advisory Committee, Heidelberg, Germany, 10-12 April 1991 (Washington, D.C.: The World Bank) (1991)

Pergamon

Adv. Space Res. Vol. 20, No. 7, pp. 1351–1360, 1997

Printed in Great Britain
0273-1177/97 $17.00 + 0.00

PII: S0273–1177(97)00729–1

TECHNOLOGICAL INNOVATIONS AND PUBLICATIONS RELATED TO SPACE SCIENCE EDUCATION

K. J. Meech

Institute for Astronomy, University of Hawaii, 2680 Woodlawn Drive, Honolulu, HI 96822, U.S.A.

ABSTRACT

As NASA Administrator Dan Goldin has noted, children are especially interested in three things: ghosts, dinosaurs, and space. We need to capitalize on the interest in space to develop and encourage a continuing understanding and curiosity about space-related issues from the grade school level through adulthood. Unfortunately, because astronomy is usually only offered in college as an introductory-level elective or as a specialized area of graduate study, most pre-college educators are uncomfortable teaching space sciences in the classroom. Additionally, the particularly dynamic nature of the field leaves educators with materials that are out-of-date and inadequate. In the framework of the new educational pedagogy focusing on hands-on activities, this paper presents a discussion of technologically innovative resources for the space science classroom (primarily at the high school level) which utilize hands-on activities for the students. The discussion will focus in five areas: major high-tech programs (which require specialized equipment in the schools), national and local workshops available to educators, hands-on activities commercially available, educational astronomical resources, and examples of hands-on activities developed and expanded upon by teachers and astronomers attending an educational workshop in Hawai'i from 1993-1995.

INTRODUCTION

With the general level of public scientific literacy declining, there is a large push to raise scientific literacy standards. This has been reflected in a White House statement issued in 1994 which stated that a goal of the United States in this area is to both produce high-quality scientists and engineers to meet the needs of the next century and to increase the level of scientific literacy in the general public by the turn of the century. An increase in science appreciation has become a national priority (Yager and Lutz, 1994) as evidenced by the national science education reform movements which include the NRC National Science Education Standards, the AAAS Project 2061 and the AGU Earth Systems Science programs, among others. Astronomy is a unique field in that it evokes a fundamental interest in exploration and the search for our origins which has long been rooted in humankind's curiosity - perhaps more so than in any other field. This public appeal, combined with the general interdisciplinary nature of the field gives it a high potential as an educational avenue for increasing scientific awareness. However, in these tight fiscal times, many educational programs are focusing on areas which can provide practical training for future job-related skills. This is particularly pronounced in areas such as Hawai'i, where recent state fiscal limitations are resulting in reductions in funding for school supplies by as much as 30%. For many public high-schools in Hawai'i, only 20-30% of the students may continue on to higher education, and districts with large percentages of minority populations have Standardized Achievement Test (SAT) scores which are below the national average. Whereas some may consider space science to be somewhat esoteric, the fascination that the field holds for young people in particular, suggests that while astronomy might not provide practical on-the-job training skills, the critical thinking skills and motivation which can be developed in students through space science and astronomy education are probably far more valuable assets than specific skills.

Astronomy and space science education in the U.S. occurs primarily in middle schools, appearing as a unit in earth sciences classes. For younger ages, while there may be some introduction to the members of the solar system, the cognitive development of elementary school children is not sufficiently advanced to understand the spatial concepts inherent with the three-dimensional nature of the field (Bishop, 1996). Unlike other sciences which get revisited in the high school years, unless there is a special interest on the part of the teacher, astronomy and space science often do not re-appear in high school. Students have their first real experience with astronomy as an introductory level course in the Universities, and a large number of students take these astronomy courses each year as a means of fulfilling distribution requirements. From this group, a few continue on to do research in the space sciences. While there is no question that the U.S. is a leader in the production of astronomy and space research scientists, we are not doing a good job at maintaining a strong science literacy in the general population. This is due in part to the fact that most astronomy is introduced rather late in the educational process, in large lecture-style classes which are not as conducive to hands-on learning. Studies have shown that the inquiry-based, hands-on learning is a far more effective concept to develop meaningful understanding for most students than the traditional lecture-based curricula.

Unfortunately, since most middle and high-school educators have themselves gone through this same system, they feel unprepared to introduce innovative space science resources into curricula at younger levels. If astronomy is taught at all, it is often limited to the identification of constellations (Slater *et al.*, 1996). Much of the problem stems from the fact that teachers have limited content background in astronomy and related sciences, with less than half the secondary teachers having had more than 6 hours of college level earth science (Weiss, 1994). Additionally, astronomy and spaces sciences are particularly dynamic fields, and those teachers who do include space science in their curricula quickly find themselves out-of-date with neither the time, resources nor the knowledge as to how to keep up in these rapidly evolving areas. There has been a recent explosion in the availability of new, innovative materials which reflect the current understanding in space science, and which incorporate the new educational standards (*e.g.* Project 2061, 1993). The purpose of this paper is to highlight sources of information on how to obtain these materials for the classroom.

The resource materials discussed in this paper were collected and developed as a result of an astronomical educational workshop for top high-school teacher-student pairs held in Hawai'i from 1993-1995. The goals of the week-long program were three-fold: (*i*) to make high-school teachers (in science and mathematics) aware of the hands-on material available and train them in the use of the materials; (*ii*) to provide students with information and what careers in science are like; and (*iii*) to create an awareness in Hawai'i of the forefront astronomical research being conducted using the Mauna Kea Observatory facilities – one of the world's premier observing sites. The workshop theme, Toward Other Planetary Systems (TOPS) was chosen in part because it is a focus of research in Hawai'i, and in part because of the interdisciplinary nature of the topic which incorporates astronomy, physics, chemistry and biology. The program was a mixture of lectures, demonstrations, hands-on activities, field trips and research. Guest speakers, totaling upwards of 30 per year, included faculty from the University of Hawai'i system, scientists and engineers from the observatories, scientists from other research organizations on the mainland, education specialists, local teachers, and graduate students. One of the great strengths of the program was the presence of the high school students which greatly facilitated the learning process for the teachers who could see first hand how the activities they tried were received by students. The students also benefited from working along side their teachers in a new capacity as fellow students. The format of the workshop placed the role of the teacher as one of facilitator and fellow student, rather than the all-knowing source of information.

RESOURCE MATERIALS

The resources described below include a mixture of high-tech materials and those which do not require advanced technology. The reality is that while many schools have modern equipment, with Internet connections to the classroom, there are many areas where resources are minimal. This latter situation is particularly relevant to developing countries, which may be in desperate need of an increase in scientific literacy but which lack any resources to bring innovative materials to the classroom. In many school systems where resources exist, there are no funds available to update and improve the materials. Another common problem which comes with educational fiscal austerity, is that class sizes are becoming larger, and the teachers are being asked to teach more classes. More than financial hardship, one of the major impediments to the introduction of space science materials into the classroom is the lack of space and time. With heavy workloads, teachers cannot be expected to seek out and locate new innovative activities, let alone develop new materials for the classroom. The purpose of this paper is to provide a concise summary of the resources available, and to prevent this listing from becoming dated before it is published, this presentation is being concurrently developed with a homepage on the World-Wide-Web, which will be continually updated as new materials become available. The resource listing may be accessed at: **http://www.ifa.hawaii.edu/~meech/education/asted.html**

In many cases the educational materials may be obtained by direct downloading from the Internet. Full information on how to obtain materials is included on the web site.

Astronomy Workshops and Hands-On Programs

There are a variety of hands-on astronomy programs which have involved significant development of curriculum materials suitable for middle school through university level education that have associated workshops for teacher training associated. Table 1 summarizes these programs with respect to level, science content, equipment / technology requirements, and cost (see also, Percy, 1996).

Hands-on-Astrophysics. This program has been developed by the American Association of Variable Star Observers (AAVSO) with funding from the National Science Foundation. The Hands-on-Astrophysics (HOA) program is based upon the unique and extensive electronic database of variable star measurements which have been made by the predominantly amateur international members of the AAVSO. The activities include measurements of known variable stars (naked eye, binocular and small telescope), followed by use of computer software to manipulate the observations along with data from the AAVSO database. Students in the program are introduced to all phases of the research process in variable star research. Materials should be available to educators in late 1996.

The Jason Project. In this program, younger students learn the principles of scientific investigation through exemplary earth science/astronomy curricula which involve students in a real scientific process. Core background materials, journals, and hands-on activities prepare the students for a project expedition which is conducted as a telepresence encounter via live TV broadcast with the collaborating scientist at the research site. Teachers are prepared for their participation in the program through development workshops which provide background, introduction to teaching standards and opportunities and instruction on use of technology in the classroom.

Hands-on-Universe. The Hands-on-Universe (HOU) curriculum materials, developed at Berkeley by C. Pennypacker, combine elements of real research into astronomy education. Astronomy labs which would have traditionally been done with the aid of small telescopes, may now be done through this program which links professional astronomers and observing facilities to the classroom. Guided through activities with a comprehensive curriculum, HOU students world-wide request observations from automated telescopes, which they download across the Internet. Participants then use HOU-developed software to reduce the astronomical images. Teachers are trained in the use of the materials through HOU educational workshops and through the use of HOU Teacher Resource Agents.

Project INSPIRE. This unique project, sponsored by the NASA Ionospheric and Atmospheric program, combines a short-term lab activity for physics students with long-term participation in a nation-wide data gathering network. For a small fee, teachers purchase a radio receiver kit (which the students build) that can detect very low frequency radio signals (*i.e.* from lightning). Specific scientifically interesting observing periods are coordinated by Bill Pine at Chaffey High School where students then can record data (with a standard tape recorder) and send the data in for analysis.

Project ASTRO. Developed by the Astronomical Society of the Pacific, this innovative program partners professional and amateur astronomers with 4-9th grade teachers to improve astronomy education and increase student interest in astronomy and science. The program helps to establish a true partnership between the three groups by providing a developmentally appropriate resource book, *The Universe at Your Fingertips*, which is a compilation of already existing hands-on materials. Additionally, the astronomers and teachers attend a training workshop to maximize the collaboration before contact with the students. The professional astronomers then make 4-10 visits to a classroom.

Image Processing for Teachers. This project educates teachers in both the technique of image processing (using a Macintosh computer) and of the content of various scientific data sets. This NSF-funded program was developed by Dr. Richard Greenberg at the University of Arizona's Department of Planetary Sciences. Teachers use the National Institute of Health software to manipulate data and to learn to teach science through scientific process. The program fee is to run a full workshop for local teachers to fully train them in the use of the software and to provide classroom materials.

FOSTER Program. The Flight Opportunities for Science Teacher Enrichment (FOSTER) program creates the opportunity for teachers to fly aboard the Kuiper Airborne Observatory (KAO). Developed in 1963, this C-141 aircraft houses a 0.9m telescope. Flying high in the Earth's atmosphere, this telescope specializes in infra-red observations impossible from the ground. Prior to flying, teachers attend summer workshops, are instructed in the use of curriculum supplement materials and the Internet. The goal is to develop a partnership between scientists and schools, and to infuse scientific method into the classroom via excitement from teacher participation. Although the KAO is being retired, its successor, SOPHIA, which will be a Boeing 747 aircraft with a 2.5m telescope, is likely to develop a similar educational outreach program concurrent with its anticipated 120 yearly missions.

Table 1 summarizes all of the above programs. The target grade level (MS = Middle School, HS = High School, Univ = University level), as well as approximate program costs, and required equipment are included. The final 7 columns indicated the focus of each program: P = Physics, A = Astronomy, E = Earth Science, C = Chemistry, M = Math, B = Biology and T = Computers and Technology.

Table 1. **HANDS-ON PROGRAMS**

Program	Inquiries	Level	Cost	Equip	P	A	E	C	M	B	T
HOA	AAVSO, Birch St., Cambridge MA 02139	HS/Univ	Free	PC/Mac binoculars		x			x		x
JASON	Jason Foundation, 395 Totten Pond Road, Waltham MA 02154	MS	Free	live TV	x	x	x			x	x
HOU	C. Pennypacker, Lawrence Berkeley Lab, B50 R232, Berkeley CA 94720	HS/Univ	Free	486PC co-proc	x	x					x
INSPIRE	B. Pine, Chaffey High School, 1245 N. Euclid Ave., Ontario CA 91762	MS/HS	$56	tape recorder	x	x	x				
ASTRO	A. Fraknoi, ASP, 390 Ashton Ave., San Francisco, CA 94112	4-9	$30	None		x					
IPT	R. Greenberg, Univ. AZ/IPT, LPL, Tucson AZ 85721	MS/HS	$900	PC	x	x	x	x	x	x	x
FOSTER	SETI Inst., 2035 Landings Drive, Mountain View, CA 94043	3-12	Free	None		x					x

Hands-on-Activities and Curricula

In addition to the extensive programs described above there have been a wide variety of excellent curriculum materials and hands-on activities developed. Most of the activities will encourage use of technology in the classroom, and will encourage students to work in collaborative settings which engage them in inquiry-based learning.

Project CLEA. Gettysburg College, with the support of the National Science Foundation, has developed a series of Contemporary Laboratory Experiences in Astronomy (CLEA), which consist of realistic observing simulations using digital images and Mac and PC based software. In this age of increasing remote observing, the software creates a very realistic open-ended research experience, where the student must make decisions about the acquisition and analysis of the simulated data similar to the process that the research astronomers go through. The level is appropriate for high school and for college level astronomy at all levels. The software is accompanied by student manuals and teachers guides, and is available at no cost by anonymous ftp (io.cc.gettysburg.edu), via the web: **http://www.gettysburg.edu/project/physics/clea/CLEAhome.html** or via U.S. mail.

Astronomy Village. The Astronomy Village is a CD-ROM based multimedia program which was developed by the NASA Classroom of the Future program at the Wheeling Jesuit College. The program was designed to supplement 9th grade earth science classes, but it is also suitable for older age groups. Using this program, students conduct astronomical investigations. While in the village, students are guided by an electronic mentor. They can listen to lectures, receive email, use the library or the computer center to process data (using the NIH image processing program) which they obtain at the village observatory. The open ended research problems mirror areas of current forefront astronomical research. This exemplary product not only introduces students to the scientific method with a wide variety of tools, but also gives the students an idea about what the life of a real astronomer is like through personal glimpses at the village cafeteria. In addition to the CD-ROM, the Wheeling Jesuit College offers no-cost teacher workshops. The Astronomy Village may be obtained at no cost through the NASA Teacher resource Centers, or through NASA CORE for a nominal shipping fee. Further information may be found at The Astronomy Village website: **http://pulsar.cotf.edu/AV/**.

Event-Based Science Modules. The Event-Based-Science (EBS) Program is an NSF-funded program developed by R. G. Wright for the Montgomery Public Schools in Maryland. The program has developed modules for middle-school earth science classes. The EBS modules are interdisciplinary in nature and rely upon the middle school student fascination with disaster in order to draw them into activities associated with a disaster, and in the process teach them science. Each module begins with video news footage of the related disaster, then the curriculum materials assist the teacher to lead the students in tackling projects simulating real-life interaction with the disaster through hands-on activities. The modules are available at a nominal cost through Addison-Wesley publishers.

Life in the Universe. The SETI Institute has developed a science curriculum project, funded by the National Science Foundation, aimed at improving the teaching of science in elementary and middle schools (grades 3-9). The project is developing teachers' guides, each of which has 10-12 science activities which focus on the possible existence of life elsewhere in the universe and is based on the experience of the SETI scientists. The guides include instructions on how to set up the activities, reproducible materials, lists of materials required and how to obtain them, lists of software, posters and slides or other materials. Classroom kits, which include a book, videotape and poster range in price from $50-$60, and can be ordered from Libraries Unlimited/Teacher Ideas Press, Dept. 9503, P.O. Box 6623, Englewood, CO 80155-6633. Sample lessions may be viewed on the web.

Project Star. Project STAR hands-on materials were developed at the Harvard Smithsonian Center for Astrophysics with funding from the NSF. High school astronomy teachers and scientists developed the curriculum materials in order to teach physical science, using astronomy as the motivation. After development, the materials were thoroughly tested and evaluated by educators and students nationwide. Materials include activities such as building a celestial sphere, building a refracting telescope, a solar system scale model, light measurement and spectroscopy etc. The materials may be obtained from Learning Technologies, Inc., 40 Cameron Avenue, Somerville, MA 02144, either individually or as teacher samplers. Those who are conducting workshops for teachers to highlight the use of the materials can obtain the kits free of charge.

Sources of Astronomy Resources Materials

Once educators become familiar with the technology that can be used in the classrooms and begin to use the available hands-on materials which have been developed, it is likely that there will be a need for additional supplementary materials. However, tracking down sources for these materials, which might include recent mission images, astronomy tools which can be integrated into existing curricula etc. can be time-consuming. Several organizations, such as the Astronomical Society of the Pacific and the SETI Institute, publish regular newletters which are free of charge to educators, which can help keep them current. Information on subscribing to these newsletters is found in the education web page described above and, in addition, hyperlinks to useful tools are found on the page.

The National Space Science Data Center (NSSDC) has had the duty of data archiving and dissemination for the NASA missions for the past 25 years. In a lesser known capacity, however, they have also played an important role in the dissemination of information to educators from elementary through high school levels. To this end, they have set up a series of Teacher Resource Centers (TRCs) at NASA Centers throughout the country. Upon written request from educators (on their school stationery) the centers will send free printed materials, and will provide videos and slides for users who submit blank media (videos/film). The TRCs are partnerships of NASA with universities and museums that serve as regional centers to provide a variety of educational materials. The Central Operations of Resources for Educators (CORE) is designed for the national and international distribution of aerospace education materials to enhance the NASA resource network. The materials available include audiovisual and curriculum materials. CORE will process teacher requests by mail for a minimal fee. For a catalog and order form write to NASA CORE, Lorain County Joint Vocational School, 15181 Route 58 South, Oberlin, OH 44074.

In addition to the NASA centers, digital images are available from many other Internet sources. A large collection is summarized in the web page being developed with this paper. A particularly useful collection of planetary images has been compiled by the NASA Planetary Data System on a CD-ROM entitled "Welcome to the Planets". This multi-media presentation highlights the best images from the NASA planetary exploration program along with associated text, audio and mission descriptions. This is available both as a CD-ROM through the NSSDC, or can be accessed directly on the web **http://pds.jpl.nasa.gov:80/planets/**. For those who do not have ready access to the technology required to access some of these materials, there is an excellent compilation of ideas and activities which have been tested in college astronomy and physics classrooms. These are compiled in a book edited by S. Pompea (1994), entitled *West's Great Ideas for Teaching Astronomy*, 2nd edition. Copies may be obtained from West Publishing, C.O.P Department, P.O. Box 64833, St. Paul, MN 55164-1803, (800) 340-9378, for $11.50 each.

National Educational Web Sites

NASA and its centers all have extensive educational resource websites which contain links to many other useful resources. Surprisingly, however, many of the national and international astronomy centers and observatories are only beginning to develop strong educational outreach programs which include information available on the Internet. These websites are listed in Table 2. There are, in addition, several web sites which contain extensive hyperlinks to sources of astronomy information on the web. A particularly thorough site is the Astroweb, and its educational resource link is shown as the last entry in Table 2.

Table 2. **ASTRONOMY EDUCATIONAL WEBSITES**

Site	Web Address
NASA Online Ed. Resources	http://www.nasa.gov/nasa_online_education.html
NASA Online Ed. Resources 2	http://quest.arc.nasa.gov/OER/
NASA Public Affairs	http://ccf.arc.nasa.gov/dx/
NASA Teacher Resource Centers	http://nssdc.gsfc.nasa.gov/planetary/ntrc.html
JPL Educational Outreach	http://learn.jpl.nasa.gov
Space Tel. Science Inst. Outreach	http://www.stsci.edu/pubinfo/edugroup/educational-activities.html
NOAO K-12 Outreach	http://www.noao.edu/education/noaoeo.html
Mt. Wilson Obs. Tel. in Ed. Pgm	http://www.mtwilson.edu/.MWO/Science/TIE/
ESO Ed. and Public Relations	http://www.eso.org/educnpubrelns/edupubrel.html
Lunar and Planetary Institute	http://cass.jsc.nasa.gov/K12.html
AAVSO Educational Programs	http://www.aavso.org/educational/programs.html
Space Science Institute Education	http://www-ssi.colorado.edu/Education/1.html
SETI Institute	http://www.seti-inst.edu/
The Planetary Society	http://planetary.org/tps/
AAS Education Initiative	http://earth.ast.smith.edu/ED/ed.html
Astroweb Educational Resources	http://fits.cv.nrao.edu/www/yp_education.html

Space Missions

With the growing national emphasis on education and science literacy, all new NASA missions are required to have a substantial educational component associated with them. Most of the past and current missions have project information and data accessible through the WWW, either via their own homepages, or through the various planetary data science nodes which serve to disseminate planetary data. There are only two missions which have engaged in extensive educational outreach development on the WWW: Galileo and the NEAR Asteroid Rendezvous Mission. The Galileo mission has online curriculum materials available and will have an online "quick response" system where students and educators can send in questions. The NEAR Mission has an excellent online resource with information about the mission, the spacecraft, and an abundance of downloadable curriculum materials and teacher information. Other missions, such as Cassini, which have been in development much longer than the fast-track Discovery class missions, have excellent web pages which clearly explain the mission details in terms accessible to the general public; however, there is no specific online educational curriculum material. The NSSDC also maintains web pages containing a wealth of information (including useful educational tools) on past, present and future missions. These may be accessed via: **http://nssdc.gsfc.nasa.gov/planetary/planetary_home.html**.

Educational Software

There exists a wide selection of astronomy-related shareware available for a variety of platforms. It would be impossible to discuss any of the software in the limited space here; however, there are several good sources which have compiled the shareware into easy-to-access formats. Network Cybernetics Corporation (**http://www.ncc.com/cdroms/as1/index.html**) has been packaging the shareware on CD rom for sale; there there are also services which summarize some of the software and make it available free of charge.

ISSAC, the Information System for Advanced Academic Computing, is a communications service funded by IBM which is operated at the University of Washington. ISAAC collects and tests a wide variety of free IBM-based software and makes it available electronically across computer networks. Researchers and educators can register for ISAAC by either by remote Internet access (telnet 128.95.32.61, and typing register at the prompt), or by submitting their name, school affiliation, mailing address (and email address, if known), and status (faculty, staff or student) to ISAAC, m/s FC-06, University of Washington, Seattle, WA 98195.

Sky & Telescope Magazine has made software available in machine-readable form that has been published in its monthly software department column since 1984. The software is written in simple BASIC, and there are nearly 100 programs available for either a Mac or DOS format as downloadable files from **http://www.skypub.com/software/software.html**. The software ranges from simple programs to calculate eclipses, ephemerides, sunset/rise times, signal to noise estimates for CCD detectors, to orbital mechanics and the computation of black hole gravity fields.

One of the goals of the Exploration in Education Initiative (ExInEd) program at the Space Telescope Science Institute is to build meaningful connections between the federally funded research astronomy programs and education of young people. To this end, they strive to make recent research results available in a useful and timely manner by publishing electronic picturebooks which are hypercard stacks (Macintosh) containing interactive information about planetary science, Earth science from orbit and space science. The stacks may be downloaded from the Internet, or may be purchased on a CD rom from either the NASA CORE office or from the Astronomical Society of the Pacific.

Funding Sources

There are sources of funding for both the professional astronomer interested in educational outreach and for the teacher who needs funds for the implementation of new activities in the classroom.

IDEA Grants. The Initiative to Develop Education through Astronomy (IDEA) program was developed by NASA in 1991 to create opportunities for the public to participate more directly in space science and astronomy. The program, which is administered through the Space Science Telescope Science Institute, offers research astronomers the opportunity to apply for public outreach grants in two categories ($6K and $20K). For information on how to apply for these grants, contact IDEA@stsci.edu, or browse the WWW IDEA site at **http://www.stsci.edu/EPA/education.html**.

NSF Programs. The National Science Foundation funds a wide variety of public outreach and educational programs through its Division of Elementary, Secondary and Informal Education (ESIE) in the Directorate for Education and Human Resources. This Division funds programs in Informal Science Education, Instructional Materials Development, Teacher Enhancement, Young Scholars and Advanced Technological Education, as well as the Presidential Awards for Excellence in Mathematics and Science Teaching. The ESIE program guidelines may be requested over the Internet at esiepubs@nsf.gov, by providing the publication number (NSF95-150), your name, address and telephone number. Forms may also be obtained by sending a request in writing to: NSF Forms and Publications Unit, 4201 Wilson Boulevard, Room P-15, Arlington, VA 22230.

Space Grant Program. The Space Grant College and Fellowship program has established Space Grant sites throughout the U.S. to develop interdisciplinary research, education and public outreach programs related to space science and engineering. As of 1995 there were 52 Space Grant Consortia involving over 550 institutions. These affiliate members gain access to fellowship and scholarship funds for students, and local educators can apply for mini-grants for the development and / or implementation of new curriculum in the classroom. Each of the Space Grant web sites has information on local contacts where educators can enquire about the availability of funding.

Other National Programs. There are several programs which are available nationally which provide funds for teacher professional development and for the implementation of new activities in mathematics and science classrooms. These include the GTE Grant program (available in 30 states) and the Tapestry Grant (concentrating on physical and environmental sciences), among others. For a complete listing or information on the various programs, contact your local Department of Education office.

CONCLUSIONS

The field of technologically innovative astronomy resource materials is growing at a tremendous rate, and there are a large number of exemplary materials available which range from complete teacher training programs, to hands-on activities and curriculum materials, to software, images and tools available over the Internet. Because the field is growing so rapidly, the description of the resources contained within this paper will be updated and maintained on the webpage discussed above.

ACKNOWLEDGEMENTS

This work was supported in part by grants from the NASA Planetary Astronomy Program (NAGW-1897), the National Science Foundation (AST-9509277), the Space Telescope Science Institute (NASA

IDEA Program; ED-90001.01-94A), and from generous support from a private donor, and corporate sponsorship from Aloha Airlines, Qantas Airlines, The Hapuna Prince Hotel, The Hawai'i Naniloa Hotel, the University of Hawai'i at Hilo Conference Center, and the Hawai'i Preparatory Academy. Special thanks is also owed all the guest speakers who donated time and creativity to help improve space science education in Hawai'i. I would also like to thank O. Hainaut for help with formatting the LaTeX for this paper.

REFERENCES

American Association for the Advancement of Science (1993). *Benchmarks for Science Literacy: Project 2061*, New York: Oxford University Press.

Bishop, J. E. (1996). "Learning Theory and Implications", in *Astronomy Education: Current Developments, Future Coordination*, Ed. J. R. Percy, ASP Conf. Series, Vol. **89**.

Percy, J. R., Ed. (1996). *Astronomy Education: Current Developments, Future Coordination.* ASP Conf. Series, Vol. **89,** Astronomical Society of the Pacific.

Pompea, S. M., Ed (1994). *Great Ideas for Teaching Astronomy.* West Pub. Co., Minneapolis. 2nd Edition.

Slater, T. F., J. R. Carpenter, and J. L. Safko (1996). "The Dynamics and Impact of a Constructivist-Based Astronomy for Teachers Course", *Jour. of Geo. Education*, **44**.

Weiss, I. R. (1994). "A Profile of Science and Mathematics Education in the United States for 1993", Chappel Hill, NC: Horizon Research (NSF RED 9150005).

Yager, R. E. and M. V. Lutz (1994). "Integrated Science: The Importance of How Versus What", *School Science and Mathematics*, **No. 94**.

Pergamon

Adv. Space Res. Vol. 20, No. 7, pp. 1361–1364, 1997

Printed in Great Britain
0273-1177/97 $17.00 + 0.00

PII: S0273–1177(97)00730–8

SMALL AIRCRAFT AS A TOOL FOR SPACE APPLICATIONS EDUCATION

C. Oprişiu*, M. I. Piso** and D. D. Prunariu**

**National Institute for Aerospace Research, Bd. Păcii 220, Bucharest, Romania*
***Romanian Space Agency, 21-25 Mendeleev str., 70168 Bucharest, Romania*

ABSTRACT

A small unmanned aircraft has been developed to experiment and teach a class of space science and technology applications. The vehicle, a lenticular aerodyne, is equipped to perform training in remote sensing, communications, attitude control and navigation. The main advantage of this system consists in the reduced period of time necessary to understand and to adjust the phases of a mission, with low costs.

INTRODUCTION

Small unmanned aircrafts are developed and manufactured to achieve low cost missions, as for civil and military survey, air photogrammetry, agriculture, and missile targeting purposes. Such vehicles are sometimes devoted to scientific purposes: as an example, the European Space Agency has undertaken the study of the potential of using a small unmanned aircraft to provide a low-cost facility for short duration microgravity experimentation by parabolic flight (Mesland *et al.*, 1995). The demonstration unit, remotely controlled by means of an umbilical cord which ensures also the telemetry, is equipped with an on-board controller which maintains the almost zero and partial microgravity levels necessary to develop scientific experiments. It is to remark that the system should include all the basic elements of a flight mission.

Another use of such systems could be realized as training and teaching tools for aerospace and space science and technology applications. Generic training courses in the field deal with specific processes which are assembled by the student. The costs of the space missions are still prohibitive to include them in direct educational processes. Online teaching in the space fields starts at the very beginning. We could mention the NASA GLOBE project for environmental studies in which students collect local ground truth data and correlate them online with the satellite imagery products. Other online interactive projects, as NASA's *Jupiter* and *HST online*, are based only on information exchange, without the possibility of direct intervention in the mission hardware.

The paper presents the progress pursued in the achieving of a small unmanned aircraft system, devoted in particular as a means to be widely used in the space applications educational process. The aircraft, developed up to the level of a prototype, is designed and constructed by the National Institute for Aerospace Researches in Bucharest. The equipment, the control centre and the mission design are produced in cooperation with the Romanian Space Agency. This system should constitute one of the basic tools in the functioning of a future Network of Space Training Centres in Central and Eastern Europe, to be developed under the auspices of the United Nations Office for Outer Space Affairs (U.N., 1996).

SMALL AIRCRAFT AND EQUIPMENT

The small aircraft is a lenticular aerodyne, with vertical take-off and landing. The lift and propulsion are assured by a duct-in propeller, the stabilization being ensured by an aerodynamic winglets setup, automatically controlled by the core inertial guidance system. Figure 1 represents the side and the cross-section views of the prototype.

The main characteristic data of the vehicle are given in Table 1 (INCAS, 1995).

Table 1. Main characteristics of the small aerodyne

Characteristic	Size
Vehicle diameter:	2 m
Propeller diameter	1.3 m
Overall height	0.4 m
Total mass	65 kg
Payload	10 kg
Maximum speed	60 km/h
Range	60 km
Flight endurance	1 h
Power plant two stroke, three cylinder engine maximum power 18 kW/7.000 rpm	

The vehicle has a command system able to control the following:

- flight at a fixed point;
- horizontal flight at various speed;
- vertical flight at various speed;
- transition from horizontal to vertical flight;
- changing of the direction of horizontal flight;
- flight (trajectory) stabilization.

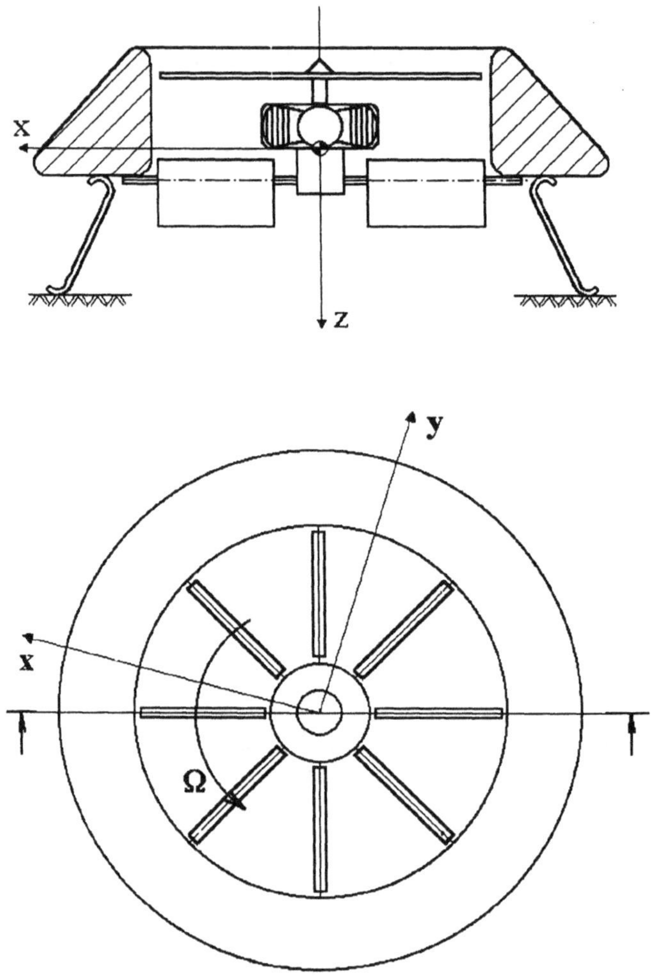

Fig. 1. Side and cross-section views of the aerodyne

The control of the vehicle is assured by a command system consisting of:

a. The sensors block:
   - three-axis linear accelerometers;
   - gyroscopic block for angular velocities;
   - height sensor;
   - airspeed sensor;
   - linear displacement of commands.

b. The onboard computer:
   - synthesizes the command laws based on the inputs received from the sensors block and the pilot/programmed output (the desired automatic evolution).

c. The electrical servocommand for the engine throttle, the four winglets for engine torque compensation and the four winglets for the attitude control.

PAYLOAD

The principal idea in establishing the payload was to create similarities between the functions of the small aircraft and those of a standard microsatellite. Following this idea, the payload is composed of remote sensing and communications equipment. Initial experiments were performed using a standard commercial TV camera as a remote sensing sensor. The communication facility has not yet been installed, because the telemetry has been ensured by means of an umbilical cord.

In the next step, the aircraft will be endowed with a matrix CCD camera, with coded digital transmission of data. Also studied is the idea to simulate a transponder, by using a S-band to X-band converter. For the better positioning of the vehicle and to supply the geocoding data, a differential GPS system is to be included on the aircraft.

MISSION OBJECTIVES AND CONTROL

The idea for the simulation of a space mission consists in the utilisation of low cost equipment with the same functions as the real one. The main tasks are:

- launch of the vehicle;
- tracking;
- control of the trajectory;
- reception of the imagery;
- primary analysis of the imagery;
- storage and the distribution of the imagery;
- recovery of the vehicle.

The system for the simulation of a launch, tracking and reception centre is composed of the ground control centre and the flying platform. A block representation of the system is given in Figure 2.

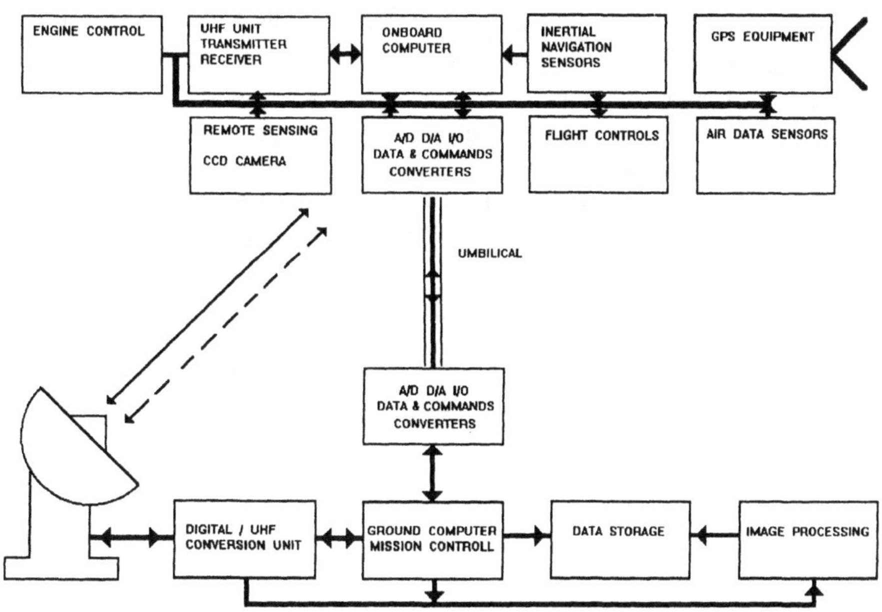

Fig. 2. Block representation of the system

The ground control centre has the following functions:

- launch and remote control of the vehicle by sending commands and receiving the flight data necessary for the flight;
- receive the imagery from the cameras;
- primary analysis of the imagery and storage ;
- distribution of the imagery.

The flying platform, including the payload, ensures the following functions:

- flight over a controlled and predetermined area (remote controlled or programmed trajectory)
- vertical take-off and landing onboard and remotely controlled operations;
- acquiring of a predetermined set of images;
- transmission of the images to the ground centre;
- stabilization (attitude control) of the flying platform.

Figure 3 depicts a view of the flying platform.

CONCLUSIONS

The small aircraft system is a model of a complete space mission, with the space and ground segments provided as their basic functions: the mission control facilities, the ground control centre, including remote sensing data processing centre. In addition, rapid validation and consideration of ground truth parameters is possible. The main advantage of this system consists in the reduced period of time necessary to understand and to adjust the phases of a mission, with low costs.

Fig. 3. The flying platform

All phases of the process should be assisted by an institutional background. Not only information and communication technology are necessary, but permanent professional staff support. These conditions may be ensured by a research or an academic institution, whose infrastructure is appropriate. The process of teaching may be extended, for example by including the methods and principles of the design of the characteristics of a predetermined vehicle, which may be a flying object, a satellite or a rocket. This could also be an exercise intending to explain the complexity of a system for Earth observation or another space mission.

REFERENCES

INCAS, *Product specification*, National Institute for Aerospace Research, Bucharest (1995).

Mesland, D., Paris, D., Huijser, R., Lammertse, P. and Postema, R., *Ballistocraft: A Novel Facility for Microgravity Research*, ESA Bulletin No. 82, 74 (1995).

United Nations, *Report of the Scientific and Technical Subcommittee of COPUOS*, A/AC.105/637 (1996).

Pergamon

Adv. Space Res. Vol. 20, No. 7, pp. 1365–1373, 1997

Printed in Great Britain
0273-1177/97 $17.00 + 0.00

PII: S0273-1177(97)00731-X

ESTABLISHING A WEB SERVER SYSTEM FOR SPACE-MAJOR STUDENT EDUCATION AND TRAINING IN MICROSATELLITE TECHNOLOGY VIA INTERNET

F. B. Hsiao, W. L. Guan, C. P. Chou and C. T. Su

Institute of Aeronautics and Astronautics, National Cheng Kung University, Tainan 70101, Taiwan, R.O.C.

ABSTRACT

Thanks to the rapid growth of Internet technology with characteristics of easy access, vast information and friendly usage, the education and training for young aerospace-related students in universities andresearch institutions have been adapted with more dynamic, fascinating, and challenging, as compared to the traditional in-class textbook and documentary teaching. Having the goal of benefiting students in learning space technology and conducting researches in microsatellite as an example, in this paper we describe the establishment of using such networking systems as Archie, E-mail, FTP, Telnet, Newsgroups, Gophers, World Wide Web (WWW), to explain our establishment of applying them for space-major student's education and training, literature survey, and information exchange and correspondence in the world. The entire system here is based on an Intel-486 PC with 8 MB RAM Linux Operating System and TCP/IP, which has proved to be feasible for teaching, economical for developing countries, and even suitable for self-study persons at home.

INTRODUCTION

BRIEF HISTORY OF INTERNET

The Internet's origins lie in a US Department of Defense project, ARPAnet, designed in the 1970s as an experiment in the construction of damage-resistant information networks. This communication system based on a logical one-to-one linking of computers via Transfer Control Protocol (TCP), addressed using the Internet Packet (IP). The basic services of Internet are to provide electronic mail (E-mail), remote access and program control of Telnet, file exchange of File Transfer Protocol (FTP), Archie, Gopher, Newsgroups reading, and Hypertext Transfer Protocol (HTTP) for the World Wide Web, and so on. Thus, in general the Internet with different types of services is a powerful tool to a student's education and training. In particular, for space-major students to access aerospace programs and organizations for relevant information via the Internet, it simply needs several seconds to minutes to complete the access. Many papers have shown the feasibility and applicability of Internet for space education and training (Johnson (1995), Orsak & Metter (1995), Welch(1994)).

MOTIVATION AND PURPOSE OF WWW SERVER IN RMRL

The Remote Piloted Vehicle (RPV) & Microsatellite Research Laboratory (RMRL) in National Cheng Kung University (NCKU) was established in 1985 with the goal to develop a cost-effective, multi-purpose platform (RPV) for studying store-and-forward communication, remote sensing, and other scientific payloads as a preliminary for microsatellite missions. To implement these goals, we candidly emphasize our major objectives on: Education & Training, New Concept/Technology Demonstration for Aerospace, cost-effective Payload to integrate with the RPV and microsatellite. In regard to the RPV design and manufacture training in RMRL, we intend to focus on the whole aircraft as a system which includes mission analysis, design processes, assembly and integration, demonstrating the flight dynamics theory, and carrying out different mission-payload experimentsAs for the microsatllite development which is a popular vehicle for space education and training, its technology is ever-growthing but still new in Taiwan. The RMRL has been carried out abundant cooperation projects with domestic/international organizations. In 1990, the RMRL successful established a microsat ground receiving station on campus, with emphasis on space education and training,satellite data communication and digital signal processing. Among many international collaborations, the SEDSAT(Wu , et al (1995)) project with the University of Alabama in Huntsville is a typical example to broaden our view ofresearch activities. To strive for future space research cooperations, more in-depth and practical study on the microsat bus and payload system shall be further enhanced. The basic hardware requirements for the RMRL server include an Intel-486 processor, 8 MB RAM, VESA bus and extension slots, internet card, etc. The PC-based Linux Server contains four major Internet services: E-mail transmission and receiving (E-mail server), file transfer site (FTP, anonymous FTP), remote assess and command control (Telnet), and WWW page (HTTP server). The server is established to be self-contained by students, and based on the purposes of space education and related technology/skill training during the implementation. Figure 1 shows the functional block diagram and Server system linkage of RMRL.

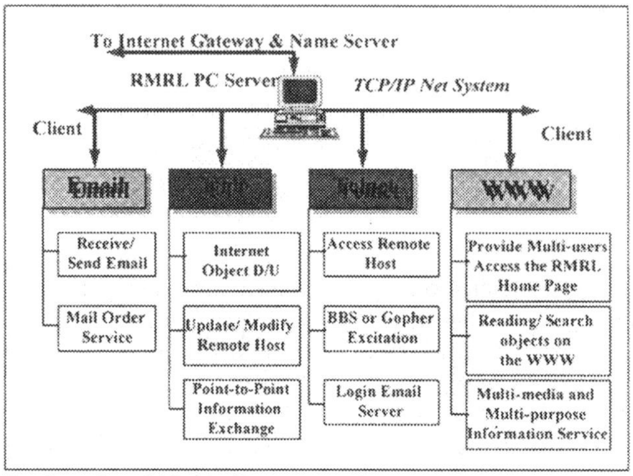

Fig. 1. The functional block diagram and Server system linkage of RMRL.

SYSTEM & FUNCTION OVERVIEW

LINUX SYSTEM

Linux is a Unix clone written from scratch by Linus Torvalds with the assistance ofa loosely-knit team of hackers across the Net. It contains almost all the features that one you would expect in a modern fully-fledged Unix, including true multi-tasking, virtual memory, shared libraries, demand loading, shared copy-on-write executables, proper memory management and TCP/IP networking.Linux runs mainly on 386/486/586-based PCs with at least 4 MB RAM and at least 10MB hard disk space required, using the Intel, Motorola or other compatible processor family hardware facilities to implement these features. Having such equipment on hand, the RMRL chose the Linux system as our multi-purpose internet server. The Linux kernel is distributed under the General Public License (GPL), and it's copyright belongs to Linus Torvalds, which means that anyone may freely copy, change and distribute it, but may not impose any restrictions on future distributions and must also make the source code available. Although the Linux is not a public domain software, it intends to encourage anyone to join the worldwide developing team for the Linux system

HOW TO ACCESS THE INTERNET INFORMATION?

ARCHIE

The Archie is a kind of mega-librarian that automatically and regularly goes out to a large amount of Internet servers and to index their files to create a single, searchable database, so that the client user can search for the text/zipped file and directory names by a fully name or partialstring. Actually, Archie is not a single system, yet it will scan Internet hosts regularly. The anonymous information index will be constantly updated to the Archie server world wide. Most Archie databases now hold information more than 2.5 million files and their related locations. The Archie client and server program files are available and free of access through the anonymous FTP at several locations on the Internet, such as **ftp.ncku.edu.tw** (/pub/winsock/archie/wsarch06.zip). These anonymous FTP's will have a common sub-directory; anyone can easilyfind and download them. The usage of anonymous FTP is a clue to one of Archie's limitations. The Archie database includes files available only in those servers enabling anonymous FTP. This means that site requires one to login. Even though that login is publicly available, it won't be included in the Archie database. Shown below is a short address list of some of the Archie servers with their geographical locations:

archie.doc.ic.ac.uk	**Canada**	archie.univ-rennes1.fr	**France**
archie-th-darmstadt.de	**Germany**	archie.unipi.it	**Italy**
archie.wide.ad.jp	**Japan**	archie.hana.nm.kr	**Korea**
archie.ncu.edu.tw	**Taiwan**	archie.unl.edu	**USA (NE)**

archie.ans.net **USA (NY)**

ELECTRONIC MAIL

The electronic mail (E-mail) is probably the most common use for any computer network. E-mail is based on the fundamental concept of store-and-forward technology. Because the E-mail uses the Simple Mail Transfer Protocol Standard (SMTP), as a result there are numerous E-mail packages running on many different platforms. Anything other than a general discussion is beyond the scope of this paper. E-mail allows swift communication between widely separated individuals. All that one needs is simply an E-mail address. There are a variety of address formats in use on different networks, but on the Internet an E-mail address has the typical form given as:

user_name@machine_name

The machine_name has two forms; one is the domain name, the other the IP address name. **Like what fbhsiao@mail.ncku.edu.tw is domain name, and fbhsiao@140.116.2.10 is IP name.**

FAQ

FAQ stands for Frequently Asked Questions. It is common in some newsgroups for particular questions to be asked frequently by a person new to or unfamiliar with the newsgroups. To avoid these kinds of annoyingly repetitive queries, many newsgroups have an FAQ list. Thus, before posting your question or troubleshooting help on the FAQ, be sure to check the similar title in the related newsgroups last posted. FAQ is usually continually updated and posted both to the newsgroups in their themselves groups and to the group**news.answers** (about once a month). Figure 2 shows an example of the FAQ resources in the Taiwan network system.

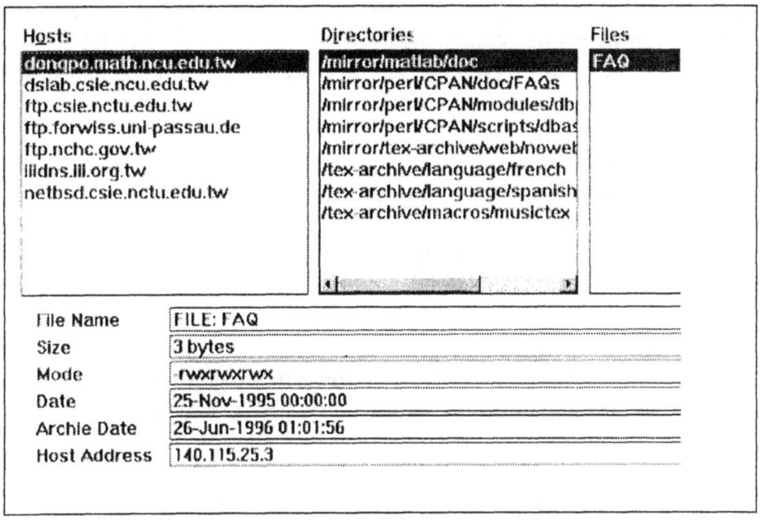

Fig. 2. The FAQ resources in the Taiwan network system.

FTP

The tool for transferring files between the computers through Internet is called File Transfer Protocol (FTP). Using a standard protocol for file transfer, FTP works regardless of the type of computer one usesor the type of machine at the other end terminal. FTP can do more than just retrieve files. One can use it to transfer files to remote host computers from a computer. To make it a practical tool, FTP includes commands for listing directories, files, and remotely operating the computer to work with the standard commands.The simplest form of FTP is called **anonymous FTP** since there is no requirement to have an account on the remote computer. The standard FTP command form is as follows:

ftp remote-computer-domain / IP name

Figure 3 shows a graphical client interface being used to access the FTP site of AMSAT-NA, in which much Public Domain (PD) software for satellite technology is illustrated.

GOPHER

Gopher is a client/server system that can be used on different platforms, including UNIX, DOS, Microsoft Win 3.1/NT/95, Macintosh, OS/2 and Nextstep. The client software runs on the terminal computer and talks to anyone of the Gopher servers. The client end of Gopher is a menu-driven program that enables one to select froma menu. It is a dynamic entity based on selections made by the user. Although the top level menu may live at the site where the gopher is, the sub-levels can be anywhere in the world. Figure 4 shows the local gopher (140.116.82.7) and its accessible Gopher server in the world.

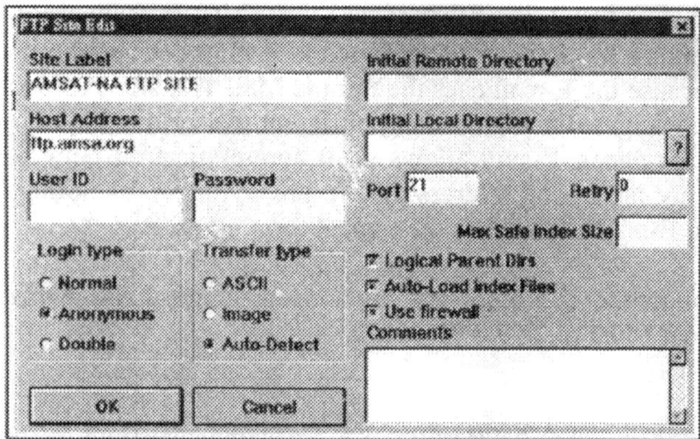

Fig. 3. A graphical client interface for FTP.

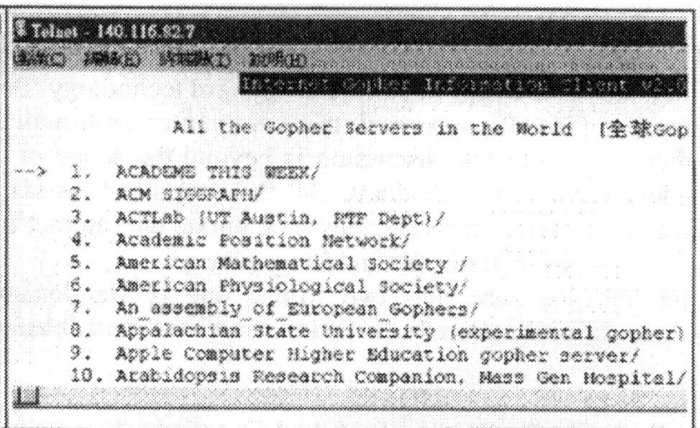

Fig. 4. The local Gopher (140.116.82.7) and its accessible Gopher Server in the world.

NEWS

News started with Usenet News developed in the late 1970s when Unix to Unix Copy (UUCP) was used as a means of distributing information between a small number of university computers in the USA. In the early 1980s after the software has been released for general usage, the number of machines using the UUCP based on news distribution increases rapidly. In the mid-1980s, this was replaced by the news software using Network News Transfer Protocol (NNTP), in which the standard TCP/IP connections (unlike UUCP) is used, and thus the number of machines connected to the news system increases drastically. Figure 5 depicts the newsgroups of **sci. space.news** and an article from which the mission of STS-78 is given.

Usenet newsgroups are divided into seven major categories dealing with different subject areas;

sci	Science and technology (but not computing)
comp	Computing
news	The news system itself
rec	Recreational activities
soc	Social and political matters
talk	Discussion
misc	Topics that do not fit any of the other categories

These areas are then split into sub-newsgroups related to the aerospace, satellite, navigation and astronomy science study. For example:

I. clari.tw.aerospace fj.rec.aerospace
II. alt.binaries.pictures.astro alt.sci.astro.aips alt.sci.astro.figaro
alt.sci.astro.fits alt.sci.image-facility alt.sci.physics.spam alt.sci.planetary
III. can.schoolnet.space.jr can.schoolnet.space.sr c-space.mkt fido7.space
sci.space.news sci.space.policy sci.space.science
sci.space.shuttle sci.space.techtalk politics.space
IV: bit.listserv.seds-l bit.listserv.sednews git.club.seds
V: rec.radio.satellite.dbs rec.video.satellite.tvro sci.geo.satellite-nav
VI: aus.radio.amsat aus.radio.packet rec.radio.amateur.equipment
rec.radio.amateur.space sci.image.processing spacelink.ask.nasa
ba.market.vehicles ucb.market.vehicles alt.binaries.pictures.vehicles

TELNET

Telnet is a program that uses the telnet protocol, part of the TCP/IP protocol suite. The remote computer, called telnet server, accepts telnet connections from a client over a TCP/IP compatible Internet system. The telnet general command form is as fellows:

telnet remote-computer-domain/IP address

There can be two different commands in use, but the same results will be obtained, as such:

telnet 140.116.2.10 [IP name address] or **telnet mail.ncku.edu.tw** [domain name address]

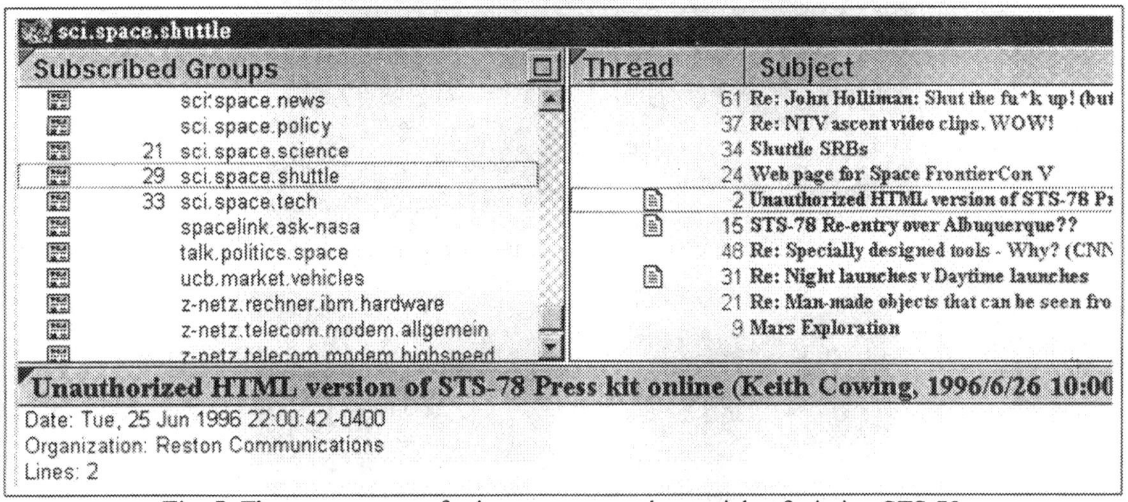

Fig. 5. The newsgroups of sci.space.news and an article ofmission STS-78.

World Wide Web & Its Browsers

The World Wide Web (WWW) is the latest information service to arrive in the Internet. It was developed by the CERN Hight Energy Physics Laboratory in Switzerland. It is hypertext document-based, distributed Internet database system, a collection of servers that offer a powerful means of accessing information over the Internet. WWW is a client/server resource as Gopher is. A text browser client available for connecting to it is called Lynx. There are many graphical clients available for it. The significant difference between the Web and the other information retrieval systems described so far is that the Web has no hierarchy. Web documents use Hypertext Markup Language (HTML) to describe their text styles and hypermedia links to image and digital sound files and to other dynamic linking documents. All of these may be on Web browsers anywhere on the Internet. The location of Web documents is given by a Universal Resource Locator (URL). This indicates the Web server that document resides on and possibly the particular page on the server as well. URL has the common form as follows:

http://www.ncku.edu.tw/index.html

in which "**http://**" stands for hyperlink protocol of WWW. "**www.ncku.edu.tw**" describes the Web document directory and server location, and the "**index.html**" the Web document file. There are a number of Web browser program available on the Internert. Anyone can use the Archie to search, and the FTP to download files. However, one particular graphical browser which is well-accepted and frequently used in the Web is called Netscape (**http://home.netscape.com/**).

SPACE INFORMATION RESOURCES OF RMRL WORLD WIDE WEB HOME PAGE

CURRENT RESEARCH TOPICS IN RMRL

The RMRL was established to develop different types of RPVs for multi-purpose applications. The RX-series RPVs were first developed for the concept demonstration and air-photographic usage between 1985-1987. Later on, the wind tunnel verification of a delta-winged RPV, named FALCON-1 and a new model offlight control interface modification and long endurance RPVs, named AEROSPACE-2000 & Thorn Bird, were successfully flight tested in the late 1980s. Since 1990, the RMRL has been developing the FALCON-series RPVs with special features of low cost, large payload capability, and multi-payload compartment design The FALCON-2 was ever invited to join the Thailand Airshow in 1994, while the FALCON-3(Hsiao, et al (1995)) was designed to integrate the CCD camera, radio transceiver, GPS receiver and on-board computer as a payload system to demonstrate the capability of near-ground surveying, real-time position navigation, and dynamic landimage transmission from the vehicle. Figure 6 shows different missions being demonstrated by various RPVs in RMRL.

Being an educational institution in NCKU, we have made great efforts to establish a receiving station for student education and training in satellite technology. By examining recent microsat missions(Sweeting (1992), Daidoff (1990)), it is noted that the investigations on radio link and transmission schemes(Ward & Price (1990), Ward & Price (1991)) are the most important topics as compared with some typical space research experiments(Hsiao & Guan (1994), Shaviv & Shachar (1994)).

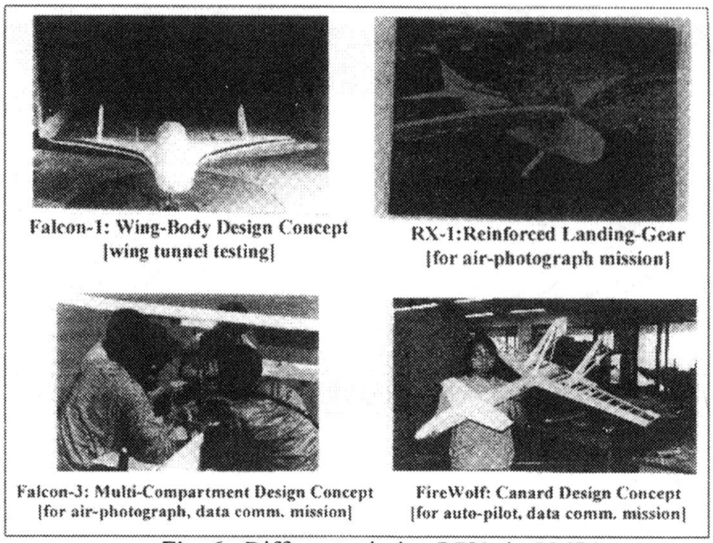

Fig. 6. Different mission RPVs in RMRL.

THE RMRL WEB INFORMATION AND RELATED SPACE RESOURCES

In order to train students via RPV system development and microsat technology, the Web server system is implemented and heavily used. Based on an Intel-486 PC with Linux system, the space information in relation to the Web site is established. The RMRL's Web site is designed to provide a rapid Internet system for access to the digital signal processing (DSP) kit, GPS navigation and positioning system, microsat technology, RPV system, and other space-related Web sites such as anonymous FTP(Hsiao & Liu (1995)) site, Ham radio agency, and satellite communication service organizations on the Internet. Figure 7 shows the Web structure of the RMRL's Home page. Sections 3.3 to 3.4 will discuss the strategy of education and training for the aerospace engineering students in IAA/NCKU.

As shown in Fig. 8, the RMRL Web pages can be used to study the avionics, control and navigation system development and multi-purpose payload design concept through sending an E-mail message rapidly to discuss with relevant persons, and searching for more information or technical papers through the Internet. The Falcon-3 RPV combines features of cost-effective, multi-purposed module design concept, on-board data acquisition and processing, and radio transmission, which has a payload capacity of 15 kg. It can be used as either a store-and-forward data/voice communication with a ground terminal, an air-photograph remote sensing, or an advanced flight navigation and trajectory system.

THE MICROSAT TECHNOLOGY-RELATED WEB INFORMATION FOR AEROSPACE STUDENTS

The Web search function and on-line library in the RMRL can be used to survey NASA teachnical reports, AIAA and Institute of Navigation (ION) publications, etc., which are now very easy to accessand download via the FTP of TCP/IP. For the course of **Space Technology Introduction & Application**, the Web Server was used to acquire the information as related to such spacecraft systems/subsystems as the Orbital Mechanics, Telemetry, Tracking & Command (TT&C), bus sub-systems, payload design concepts verification, and the like Regarding the education and training of students, the access to anonymous FTP sites to download several orbital elements (orbit parameters) for the satellite tracking are very important. Examples of the Windows Satellite Program (WISP), Space Shuttle Tracking Software (STS), and Satellite Navigation Program (Satnav) are among some of PD software. Typical microsat activities are at the University of Surrey (UoSAT, UK) and the Weber State University (WeberSat, USA). As for non-profit agencies, the hamradio group (Amateur) is the best example which has developed many Amateur microsatellites such as the OSCAR-series, Phase-3A~3D, Fuji-OSCAR-12/20, etc. The next generation micro- to nano-class satelliotes will carry out multi-purpose communication services for mobile terminals by constellations in LEO to HEO, to investigate the environment, remote sensing and functional/stratum management for security and personal navigation services. Inthese days, the applications of GPS are many and so the RMRL placed emphasis on investigating the GPS navigation system aboard an RPV. The main objective is to study the autonomous orbital navigation using GPS data for LEO microsat trajectory navigation and tracking. Thus, the Web information about the GPS, DSP, and the data communication system will be more efficienctly surveyed via the WWW browsers and any search engine, such as **Alta vista, Infoseek ,YAHOO,** etc.

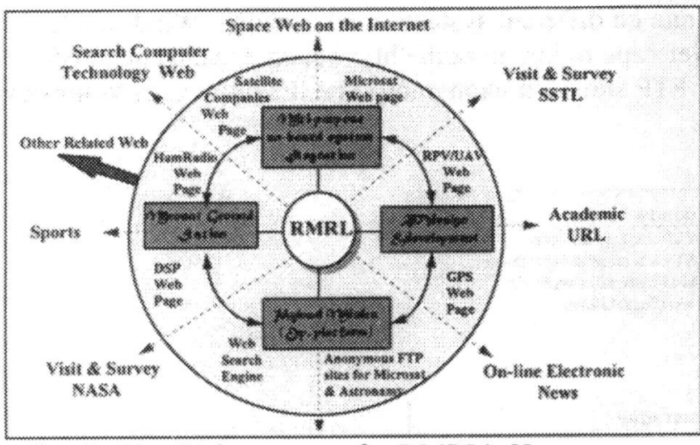

Fig. 7. The Web structure for RMRL's Home page.

Fig. 8. The Home page information of RMRL.

CASE STUDY & DISCUSSION

SEARCH AND DOWNLOAD MICROSAT TRACKING PROGRAM

In this section we describe how to search a satellite tracking program using Archie and how to use FTP of TCP/IP to download some interested files in an anonymous FTP site. From the interactive menu, one should key in the search string "**logsat**" and then to get the file as needed from the several anonymous FTP sites. The next step isto de-compress and install the logsat in Windows step by step. Figure 9 lists the detailed proceduresfor download to logsat30.zip from nctuccca.edu.tw.

AN EXAMPLE OF VISIT TO SSTL AND DOWNLOAD *.PDF FILES

Using WWW browser-Netscape via the HTTP service connecting to the home page of Surrey University, an example of literature survey about the SSTL research activities on remote sensing for microsatellites is presented. Because most of the client browser programs only support on-line display and dynamic hypertext document connections, how to keep the desired information in the client terminal is very important for database management. In general, some home page information is compressed to the hypertext document as postscript format (**pdf**). Hence, using the FTP or HTTP to download the postscript files will be more efficient and easy to take post-processing of the document. First, one has to get the URL of SSTL from some space-related magazine or to conduct a search from the HTTP searching engine, such as YAHOO. Secondly, one starts to run the WWW browser-Netscape and to input the URL as **"http://www.ee.surrey.ac.uk/EE/CSER/UOSAT/index.html"**, showing on PC screen the interested documents. Finally, suppose one expects to select the **"xian95.html"**, then continue to read this paper and save as a hypertext format file at a client terminal. If one wants to have direct editing of this document,the HTML program-HTMLabc can be used. Figure 10 shows the procedures of HTTP link to CSER, hypertext document download, and postscript data processing.

A CASE OF VISIT TO RMRL WWW & FTP SITES

In this section we discuss how to setup the Internet system in order to run WWW (client) browsers program and to use the FTP program to download AMSAT or microsat related PD softwares from RMRL or related anonymous sites. As shown in Fig. 11, the procedures for network system initization and setup, WWW browser progrom, and FTP program setup are sequentially explained as follows:

There are three major procedures for operating the FTP and HTTP:

1. Network system initization and setup: This means that one needs to setup the netcad interrupt number, vector number, and to configure the IP, name server, gateway, and netmark.
2. WWW (client) browsers program: There are many browser programs for which anyone can use through the Archie searching. "**Netscape**" is one of the best browsers available, which can also support the JAVA function. The URL of RMRL is "http://140.116.200.187/". Since RMRL has only one library in IAA, NCKU, one can also pay a visit to IAA and NCKU. The URL's are "http://www.iaa.ncku.edu.tw" and "http://www.ncku.edu.tw", respectively.

3. FTP program: To use FTP is very easy, but it can operate on different systems, such as DOS, Windows 3.1/95, etc. One of the most efficient methods is to use the Netscape to key in both "http://ctaxp2.iaa.iaa.ncku.edu.tw" and "http://ctaxp2.iaa.iaa.ncku.edu.tw/". The RMRL's FTP site is an anonymous site. Its purpose is to improve the time and cost for remotely downloading softwares.

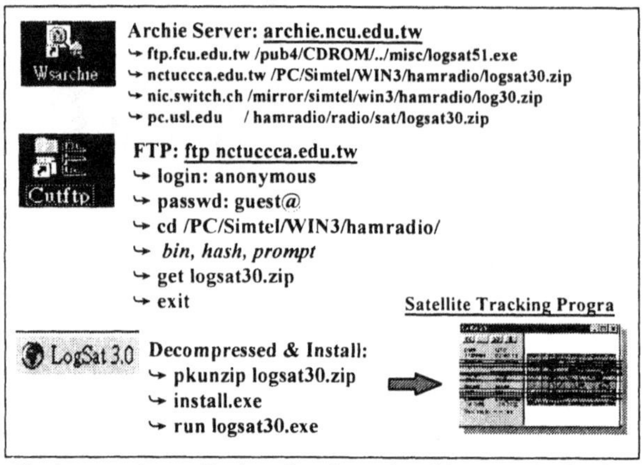

Fig. 9. Detailed procedures listing for download logsat30.zip from nctuccca.edu.tw.

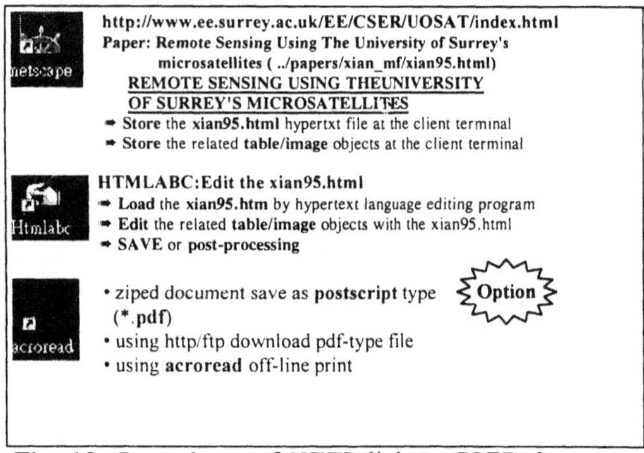

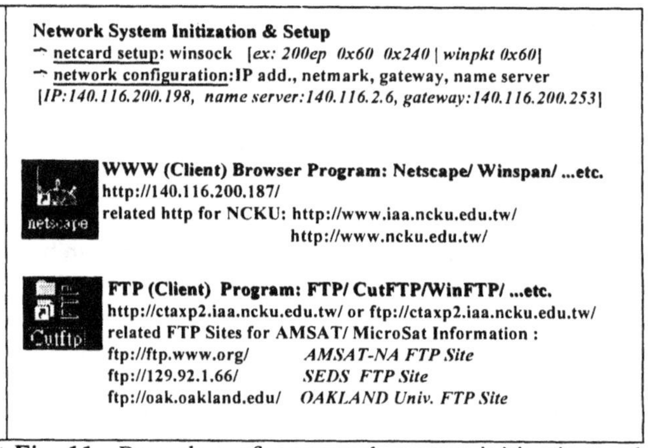

Fig. 10. Procedures of HTTP link to CSER, hypertext document download, and postscript data processing.

Fig. 11. Procedures for network system initization and setup, WWW browser progrom and FTP program setup.

CONCLUSION

This paper has provided a general introduction to the space-related techniques and tools available for accessing information via the Internet in the home page of the RPV and Microsat Research Laboratory (RMRL), IAA/NCKU. The home page information offers several pointers to sources and types of information of particular interest to space major students and educators. We have established a PC-based Linux server to be used for different kinds of Internet service. It has also been tested to be a good tool to conduct search, read, and interactive message exchange of aerospace-related information via the RMRL home page and the relevant URLs.

ACKNOWLEDGEMENT

This work is supported by the National Science Council, R.O.C., under Contract No. NSC 83-NSPO-A-RDD-006-30.

References

Daidoff, M., "The Satellite Experimenter's Handbook", The American Radio Relay League, pp.5.1-5.6, 1990.
Hsiao, F.B. & Guan, W.L., "Ground-based Simulation for microsatellites Telemetry Transmission and Image Data Processing", 45th Congress of The International Astronautical Federation, IAA-94-744, 1994.

Hsiao, F.B., Guan, W.L., Cheng, S.W. and Liu, M.C., "The Study of Telemetry Data Processing and Error Bit Detection for Microsatellite Communications", 46th Congress of The International Astronautical Federation, IAA-95-B.6.10, 1995.

Hsiao, F.B. and Liu, M.C., "The Design and Applications of Remotely Piloted Vehicles", The 4th conference on National Defense Technology, Chungli, Taiwan, Nov. 1995.

Johnson, D.N., "Signal Processing and the World Wide Web", IEEE Signal Processing Magazine, 1053-5888/95, IEEE, pp.53-57.

Orsak, G.C. & Metter, D.M., "Collaborative SP Education Using the Internet and Matlab", IEEE Signal Processing Magazine, 1053-5888/95, 1995IEEE, pp.23-32.

Shaviv, G. & Shachar, M., "The TechSAT-1: An Earth-Point, Three-Axis Stabilized Microsatellite", 45th Congress of The International Astronautical Federation, IAA-94-IAA.11.2.767, 1994.

Sweeting, M.N., "Uosat Microsatellite Missions", Electionics & Communication Engineering, pp.141-150, 1992.

Ward, J.W. & Price, H.E., "PACSAT Protcol Suite - an Overview", 9th Computer Networkin Conference, 1990, pp. 203-255.

Ward, J.W. & Price, H.E., "Protocol for Store-and-forward Message Switching via Microsatellites", 5th Annual AIAA/USU Conference on Small Satellites, 1991, pp.1-19.

Welch, C.S., "Internet Electronic Data Results for Space Students", 45th Congress of The International Astronautical Federation, IAA-94-1032, 1994.

Wu, S.T., Wingo, D., BankSton, C. and Hsiao, F.B., "An Overview of the Joint UAH-NCKU SEDSAT1/SEASIS Program", Oceanic Microwave Remote Sensing Workshop, Univ. of Delaware, Aug. 16-18, 1995.

Pergamon

Adv. Space Res. Vol. 20, No. 7, pp. 1375–1378, 1997

Printed in Great Britain
0273-1177/97 $17.00 + 0.00

PII: S0273-1177(97)00732-1

POSTGRADUATE TRAINING FOR SPACE SCIENCE AND TECHNOLOGY EDUCATION

D. Andreescu*, M. I. Piso* and M. Niţă**

*Romanian Space Agency, 21–25 Mendeleev str., 70168 Bucharest, Romania
**Department of Aeronautics, "Polytechnica" University Bucharest, Romania

ABSTRACT

The paper describes an experimental form of multidisciplinary postgraduate training in space science and technology, developed under the auspices of the Romanian Space Agency in the period 1994 - 1996. This specific form of training is recommended for developing countries.

INTRODUCTION

Space means globalization and integration. The access to space opportunities means, first of all, the admission as an active participant to the efforts to maintain and develop human civilization.

The endeavour of the space era has been performed mainly by major states. Public funding, which constituted the main support, has been justified due to the states' specific needs of national security and image. The tremendous effort in the field of space science and technology generated a growing advance of those states versus the others, in particular in the level and degree of specialization of an increasing number of space professionals. This trend generated a growing gap between the space powers and the other states. To develop space technologies, specialized personnel is needed. The lack of expertise in the non-space countries is the essential difficulty in an attempt to make use of the space technology, that is necessary for development.

Two major aspects should be emphasized:

a. The successful transfer of space technology to a region or a country requires, first, the sufficient qualified personnel in science, technology and management. The utilization of foreign experts is mostly prohibitive due to the higher costs - the costs of salaries and local expenses being larger as those demanded in their own countries.

b. Most universities and other higher education institutions, in particular from developing countries, generate as output professionals in restricted fields. The rate of evolution of space science and technology is higher than in traditional academic domains.

Global view. Multi- and inter-disciplinarity

It is important to point out that space science and technology are not essentially different from "terrestrial" science and technology. Outer space is, in a way, a larger environment in which our home planet is embedded, and the physical laws that determine the natural phenomena therefore the same as on Earth. However, the scale and some specific physical conditions are different. In this way, all scientific models describing the outer space phenomena, together with the relations between them, should be corrected to fit the same surrounding conditions. A professional, in order to extend the common "terrestrial" thinking to accommodate it to the space environment, should make use of some "global" concepts of investigation and thinking. Those concepts are mostly employed in basic and advanced science. Recent documents endorsed by the United Nations Committee on Peaceful Uses of Outer Space include the recommendation that ". . . in promoting spin-off benefits and effective space applications, particularly in the case of developing countries, it was important that the capacity to understand the technology and to develop it were considered to be of primary importance. The Committee was of the view that developing countries should develop their potential in basic and advanced research in order to promote capacity building for space applications and to benefit from spin-offs of space technology" (U.N., 1996b).

It is not easy to create professionals in global thinking, their formation needing much background in research and applications, together with international ambience, awareness and practice. Such specialists are required for strategic management. However, a good procedure to improve capacity in global space understanding is the promotion of multi- and inter-disciplinarity in education, training and research. Professional specialization in two or more apparently disjointed fields belonging to the traditional academic disciplines (such as basic theoretical and experimental research in physics, mechanics, aeronautics, information technology, applied research, law, engineering) generates the capacity to build short and long range connections between the models describing different parts of space scientific, technical or legal characteristic processes.

Major space educational initiatives

In the last decade, the rapid development of space activities, in particular space applications, generated particular forms of multi- and interdisciplinary training (see U.N., 1993b). A common feature of these activities is the wide international structure of the students and the professorial staff. Among those initiatives, we may cite the *International Space University*, (ISU, 1995), developing ten week summer and eleven months masters courses. The *Core curriculum*, containing: Space Architecture, Space Business and Management, Space Engineering, Space Life Science, Space Policy and Law, Space Resources and Manufacturing, Satellite Applications, Space Physical Sciences, Space Humanities, Space Information, is followed by an *Advanced curriculum* and a *Design project*. Over a hundred postgraduates from all continents are trained at ISU each year. There are also other significant international establishments, such as the *International Institute for Aerospace and Earth Sciences* in Netherlands (ITC, 1994) and *Le Groupement pour le Dévelopement de la Télédetection Aérospatiale*, Toulouse (GDTA, 1994).

An important initiative has been taken at the level of the United Nations Office for Outer Space Affairs which, in response to the recommendations of the UNISPACE '82 Conference, developed in 1990 a proposal for the establishment of *Regional Centres for Space Science and Technology Education*, in particular for the developing countries. The decision was taken in the view that "... an essential pre-requisite to a successful space technology applications is the building of various essential indigenous capacities, particularly human resources, within each region." (U.N., 1990). A comprehensive project document was issued in 1993 (U.N. 1993a). This initiative is in progress for some U.N. regions, the most recent project concerning an educational system consisting of a network of space science and technology education institutions, for the region of Central, Eastern and South-Eastern Europe (U.N., 1996b).

DESCRIPTION OF THE TRAINING COURSE

An experimental form of multi-disciplinary postgraduate training in space science and technology has been developed in 1994 - 1995 under the auspices and financial support of the Romanian Space Agency (ROSA). The public announcement of the course has been followed by the selection of twenty students. A university degree and an age less than forty years were required. The selection criteria were based on the curriculum vitae, a form completed by the participants and an interview with the executive staff of ROSA. The teaching process is contained in three phases, each with a duration of 8-10 weeks.

Phase A contained of ten courses, each of 4 hours, twice a week, in space topics. The lecturers were selected from distinguished personalities of the Romanian space system, as university professors, scientific heads of research institutions, officials, with a pertinent general view on the specific space fields: astrophysics, geophysics, microgravity, life sciences, telecommunications and information, Earth observation and remote sensing technology and applications, aerospace structures and flight dynamics, space policy and law. The Phase A topics and the description of the lecturers are given in Table 1.

Table 1. Topics of the Phase A course

Number of hours	Topics	Lecturer / Scientific Title/Position
4 h	*Space activities across the world. The Romanian Space Agency and Programme. International Relations*	Dumitru ANDREESCU, vicepresident of ROSA
3 h	*Astronautics and space flight*	Dumitru PRUNARIU, astronaut
3 h	*Space vehicles and space transportation*	Augustin PETRE, Professor, Department of Aerospace Engineering, Polytechnica University Bucharest
3 h	*Microgravity and gravitational researches*	Marius PISO, Head of the Gravitational Researches Laboratory, Bucharest
3 h	*Space Dynamics*	Mihai NIȚĂ, Professor, Department of Aerospace Engineering, Polytechnica University Bucharest
3 h	*Space Medicine and Biology*	Maricel ULUITU, Director of the Institute on Normal and Pathological Physiology Bucharest
3 h	*Astrophysics and Cosmic Rays*	Dumitru HAŞEGAN, Director of the Institute on Gravitation and Space Sciences
3 h	*Space Astronomy*	Magdalena STAVINSCHI, Director of the Astronomical Institute of the Romanian Academy
3 h	*Space communications and Information*	Alexandru SPĂTARU, Professor, Director of the National Centre on Space Communications and Information
4 h	*Remote sensing and Earth Observation*	Nicolaie OPRESCU, Professor, Head of the Remote Sensing Laboratory, UTCB Bucharest

The courses were held in the evenings at the site of ROSA. A book including abstracts of the courses topics was provided to all participants. The average age was 29. The initial specializations of the students were as follows: Mechanical and Civil Engineers - 4, Electric Engineers - 3, Aerospace Engineers - 3, Research Physicists - 4, High School Teachers - 3, 1 Priest, 1 Lawyer, 1 Air Force officer.
Phase B has been structured on in-place applications and experiments, developed in relevant research and higher education institutions (Table 3). Senior scientists performed general overviews of the institution's activities and guided the students in visiting the appropriate aerospace and information facilities.

Table 2. Host institutions for Phase B Courses

Number of sessions	Host institution
1	Astronomical Institute of the Romanian Academy
2	National Institute for Aerospace Researches
1	Institute of Geodynamics of the Romanian Academy
1	National Centre for Space Communications and Information
1	Intelsat Ground Station - Cheia
1	Institute of Normal and Pathological Physiology
1	Institute on Aeronautical Medicine
2	Remote Sensing Laboratory - Technical University for Civil Engineering
1	Institute of Optoelectronics - Remote Sensing Department
1	Romanian Space Agency

In Phase C, in progress, the students are developing project works, under the supervision of the lecturers from the previous phases. The subjects, presented in Table 3, were proposed by the candidates, according to some particular goals of the Romanian Space Programme.

Table 3. Selected topics of the project works

Optical sensors suitable for satellite remote sensing uses
Applications of information technology to a public library
Space topics common for science and theology - comparative study
Guidance and attitude control onboard systems
Fluid science microgravity experiments
Optimal small satellite low orbits for remote sensing applications

After the end of Phase A, the participants received a certificate of completion. Phase C will be ended with a presentation / examination of the project by the staff of the Romanian Space Agency.

CONCLUSIONS

The general character of the course and the relative ease of organizing it may provide for a national space agency the possibility to select future space professionals, officers and educators. This form of training is suggested as a low-cost way to improve capacity building in the space fields for developing and Eastern Europe countries.

REFERENCES

International Space University, *Summer Session Program*, and *Master of Space Studies (MSS) Program* (ISU 1995).
International Institute for Aerospace Survey and Earth Sciences, *General Information*, *Land Information Survey*, *Advisory Service*, ITC Entschede 1990-1992 (ITC 1992).
Groupement pour le Developement de la Teledetection Aerospatielle, *Formation*, Toulouse (GDTA 1992).
United Nations, *Report of the Committee on the Peaceful Uses of Outer Space*, A/AC.105/456 (1990).
United Nations, *Centres for Space Science and Technology Education*, A/AC.105/534, New York (1993a).
United Nations, *Education, training, research and Fellowship Opportunities in Space Science and Technology and its Applications*, A/AC.105/542, United Nations Litho 33027, New York (1993b)
United Nations, *Report of the Scientific and Technical Subcommittee of COPUOS*, A/AC.105/637 (1996a).
United Nations , *Draft Report of the Committee on Peaceful Uses of Outer Space*, A/AC.105/L210, 1996b).

Pergamon

Adv. Space Res. Vol. 20, No. 7, pp. 1379–1382, 1997

Printed in Great Britain
0273-1177/97 $17.00 + 0.00

PII: S0273-1177(97)00733-3

USING SPACE SCIENCE AND TECHNOLOGY AS AN EDUCATIONAL TOOL: TWO DIFFERENT APPROACHES

S. K. Dunkin*, A. J. Ball** and E. A. Taylor**

**Department of Physics and Astronomy, University College London, London, U.K.*
***Unit for Space Sciences and Astrophysics, University of Kent at Canterbury, Canterbury, U.K.*

ABSTRACT

Successful ways of using space science and technology as a vehicle for science education are explored. We report on activities centred on two educational forums - "extra-curricular", in the form of special lectures delivered to over 150 pupils from different schools in one session, and "curricular" activities where the scientist entered a school for several days to become an integral part of the pupils' class. Space science and technology were applied to the school context within the framework of the UK's National Curriculum for ages 13-17. These activities were part of the Pupil Researcher Initiative, run by Sheffield Hallam University on behalf of the Particle Physics and Astronomy Research Council (PPARC), and the ESSTeL lectures sponsored by PPARC's Public Understanding of Science and Technology initiative. We give an account of the preparation of both initiatives on the part of the teacher and give an insight into the practicalities of organising the events. We comment on the advantages of using space as an educational tool and its ability to motivate the target audience.

INTRODUCTION

The fields of astronomy, space science and technology have long had the ability to pass on a sense of wonder to all ages, from primary level children to the oldest of us alive. One of the great advantages of these fields is that they are open to all; everyone can look up into the night sky and see the Moon and the stars, everyone can read about the latest space science missions being launched into space, and everyone uses the high technology brought about by the space age - satellite TV, telecommunications and weather forecasts. This accessibility to the general public and to schoolchildren can be used to put across the importance of science, maths and technology subjects, by showing the relevance to the average "person-in-the-street".

Over the years, many talks and lectures have been given to schools and public audiences about space science and technology. In 1995, two schemes were independently set up to use these subjects actively to encourage school-age children to opt for science and technology subjects after they reach the end of the first stage of their schooling. In England, all students up to the age of 16 must study science and maths, side by side with other subjects. But after that age, when they choose A-level studies, they are free to drop all science based subjects should they choose to do so. Many students find the sciences, particularly physics (essential to space science and technology), very difficult and often drop the subject from their studies to opt for something "easier" or "more interesting". These two schemes hope to show the importance of science in our everyday lives, and to show students that science has applications and benefits in careers that they were unaware of.

THE PUPIL RESEARCHER INITIATIVE: PRI

The Pupil Researcher Initiative (PRI) is a major initiative from two of the UK's science research councils, EPSRC (Engineering & Physical Sciences Research Council) and PPARC (Particle Physics & Astronomy Research Council). The PRI is coordinated by Sheffield Hallam University and, after an initial pilot phase, is a now a nationwide programme. Its aims are as follows:

* To encourage the motivation and achievement of young people in science
* To develop scientific skills and knowledge through investigations and IT
* To provide support for teachers
* To promote the public understanding of science and research
* To use up to date contexts and frontier science
* To reflect realistic science and engineering methods
* To mobilise the research community

The main activities of the PRI are as follows:

* Pupil Research Briefs
* Researchers in Residence
* School Science Fairs and Roadshows
* Pupil Researcher Journal and Conferences
* Teacher Bursaries for placements in research laboratories
* "Research School" awards
* In-Service Support for science teachers

Of these activities it is the Researcher in Residence scheme which we consider here. The scheme connects individual science Ph.D. students with a school in their local area. The student obtains payment and expenses for a total of four days visits to the school, during which he / she acts as a "researcher in Residence", assisting the school's curriculum activities at Key Stage 4 (ages 15-16), though activities are not restricted to this Key Stage. Participating students attend a training day, organised by their regional PRI coordinator. Once students and participating schools have been "matched" by the regional coordinator, the student arranges an initial visit to the school to discuss the approach to be taken and possible activities during the student's visits.

In the case of Ph.D. students working in space-related fields, the PRI provides an opportunity to visit a local school and use space as a vehicle to enhance the pupils' curricular activities at Key Stage 4. Helping at a school often proves to be a very rewarding experience for the student, who has the chance to develop communication skills for a broader audience.

Possibilities for activities as part of the scheme include the following:-

* Talks on current research in the student's institute, or any other space-related topic with which the student is familiar
* Workshop activities on a space-related topic, for example construction of a model satellite or space station
* Help with a long-term project related to space

* Short talks targeted to specific topics in the curriculum (using space to illustrate a topic)
* Visits of teachers & pupils to the student's institute
* Use of the institute's information resources to assist the school
* Participation of the school in national PRI activities such as conferences or journals
* Advice for sixth formers considering science careers

It is important to note that the Researcher in Residence is never left in sole charge of a class of pupils. The teacher always remains in overall charge.

EDUCATIONAL SPACE SCIENCE AND TECHNOLOGY LECTURES: ESSTeL

The second of the two schemes, and by far the more modest is ESSTeL, an initiative formed in 1995 and operated by two alumni of the International Space University summer session in Stockholm, Sweden. Sponsored by the Particle Physics and Astronomy Research Council (PPARC) under their Public Understanding of Science scheme, ESSTeL delivers space science lectures to schoolchildren around the country. What is different about ESSTeL? Firstly, lectures are delivered by two speakers selected from a list of young space professionals (including post-graduate students) held by ESSTeL. Lectures are given to audiences of 150+ 13-17 year old students. To achieve this number, three or more schools usually bring 50 or so students each, and to accommodate this number, a school hall or other local venue is normally used.

Each 45-60 minute lecture holds an enormous amount of information, far more than the audience are expected to take in. This is deliberate; the pace of the lecture is fast, active and hits the audience with facts and information at an incredible rate. The audience is given a sense of wonder and awe of the field of space science before being brought "down to Earth" by explaining that they are already learning the basics of all that they have seen during the talk. The closing part of the talks are specifically aimed to give the student a realisation of the vast number and variety of careers related to space science and the feeling that these careers are not only for the "elite" students, but that some part can be played by all abilities in the space arena.

LOGISTICS OF THE EVENT

The format of the ESSTeL lectures generally mean that the teacher has to spend some time organising the large number of students, and arranging a mutually agreeable time for the lecture. If the lecture is being held in a school, that teacher must arrange suitable presentation equipment, whilst the teachers from the other schools must arrange the excursion for the students attending the lecture - not always an easy task. Thus far, we have received no complaints as to the format, and all who have attended ESSTeL lectures have benefited from the experience. To aid the teachers in following up on the lecture, information packs, including a full transcript of the talk are distributed to the schools.

Schools participating in the PRI Researcher in Residence scheme are matched with Ph.D. students from a nearby university by the PRI's Regional Coordinator. The school must appoint a particular teacher to act as the contact point for the student, who will already have attended a training day. The teacher and student then need to decide on a mutually acceptable date and time for the first visit. During this visit the student and teacher discuss the general format of activities during the four days available. If the student is to participate in a long-term project it is clearly important that a timetable be laid down in advance. The eventual programme of activities is based on a combination of ideas from both the teacher and the student.

A degree of preparation is often necessary on the part of the student as well as the teacher, though basic talks or workshop sessions may require nothing more than a blackboard or slide projector.

SPACE AS A MOTIVATOR?

To use space as a motivator, there must be topics of interest to all ages and abilities. We have found that space does indeed have topics of interest to virtually all students. Using this fact, ESSTeL have found that students are attentive during the whole lecture, an indication that they are interested by what is being said. The quality and depth of some of the questions received show an understanding of the subject matter and an urge to know more. Many have studied news reports of topical issues and commented on those; some have even gone on to read books and articles on specific subjects touched upon by the lectures. Feedback from the teachers themselves has been very positive, with yearly follow-up lectures being arranged. One disadvantage we have found is that some students tend to back away from the space science because they view it as a "high-brow" career, something not for them. We try to dispel this notion, and we feel that we do surprise the students with the number and variety of careers available to them in space science. Overall, we feel that ESSTeL has been a success in its aim at motivating young students in its first year of operation. Whether this will result in some of the audience taking science subjects after age 16 remains to be seen.

Space can be used to enhance the coverage of science subjects, being both an integral part of the National Curriculum as well as a complementary topic to illustrate a wide range of scientific concepts. Pupils at schools participating in the PRI Researcher in Residence scheme can benefit from visits by a space-related Ph.D. student. Pupils often enjoy the change from regular activities provided by a visit. The space field is a valuable source of inspiring and unusual images and information which can capture pupils' attention and interest, as well as stimulate further questions and thoughts which might be followed up later within the school or during a subsequent visit. The continuing nature of the PRI connection is one of its particular strengths.

In conclusion, we have found that space is a unique and invaluable vehicle with which to relate scientific concepts to children. Space holds a fascination for us all, and when used to complement regular teaching can be inspiring and enlightening to the young people whom we are trying to reach.

ACKNOWLEDGEMENTS

SKD and EAT would like to acknowledge the support of the Particle Physics and Astronomy Research Council, particularly those associated with the Public Understanding of Science and Technology Department. We would also like to thank Mrs. Sue Bayford of the Space Education Trust for the administrative support which she has provided. AJB would like to thank Andy Piggott of the Royal Institution for the co-ordination of the PRI scheme in the London and SE areas, and Becky Parker, Head of Science at the Simon Langton Girls' School, Canterbury.

REFERENCES

Educational Space Science and Technology Lectures, http://www.gbnet.net/orgs/esstel/
Particle Physics and Astronomy Research Council, http://www.pparc.co.uk/
Particle Physics and Astronomy Research Council, *The Pupil Researcher Initiative (brochure)*, (1995)
The Pupil Researcher Initiative, http://www.ri.ac.uk/Education/PRI/PRInet.html

Pergamon

Adv. Space Res. Vol. 20, No. 7, pp. 1383–1386, 1997

Printed in Great Britain
0273-1177/97 $17.00 + 0.00

PII: S0273–1177(97)00734–5

SPACE PLASMA EDUCATION AT MOSCOW STATE UNIVERSITY

E. E. Antonova

Skobeltskyn Institute of Nuclear Physics, Moscow State University, Moscow, 119899, Russia

ABSTRACT

The beginning of energetic particles and space plasma research with the launching of the first satellites brought the necessity of educating specialists in this field. In Moscow State University such specialists are educated in the Cosmic Rays and Space Physics Group. For students who intend to participate in satellite experiments this education includes a Space Plasma lecture course. This course includes the of main space plasma problems such as theoretical approaches to plasma description (single particle, fluid and kinetic approximations, plasma waves and instabilities, weak and strong turbulence, solitons and shock waves, etc.) and the main aspects of solar atmosphere physics, solar wind, magnetosphere and ionosphere of the Earth. The main goal of the education process is to prepare a student for independent scientific research which generally begins in the sixth semester. In most cases the student's diploma work is followed by a publication in one of the space journals. The main problems of increasing the quality of the education are analyzed.

INTRODUCTION

The tuition of specialists in space researches in the Cosmic Rays and Space Physics Group of the Physics Faculty of Moscow State University is a complex process which includes theoretical and experimental training. It begins in the third year of education after the students have chosen their future specialty and been assigned to different Groups. Historically the Cosmic Rays and Space Physics Group is a part of the Nuclear Department of the Physics Faculty since the first researches of cosmic rays in Russia were connected with the solution of the problem of nuclear interactions of high energy particles with matter and were conducted on balloons. New possibilities of cosmic ray researches on rockets and the first Soviet satellites led to some changes in the Group's orientation. One of the main aspects of this reorientation after the discovery of the radiation belts became the physics of the near-Earth cosmic space and heliosphere. Since matter in these regions is mainly in the plasma state, one of the main aspects of space research are now plasma processes in cosmic space. In the current paper we describe the main features of student education in the field of space plasma physics and try to stress the main problems which arise in the process of such education.

THE PROBLEM OF CHOICE

One of the main problems which the student meets during the third year of his (her) education is the choice of the Group. In order to help the student with this decision we organize special seminars and meetings with famous scientists working in the Group. Students have the possibility to visit the laboratories of the Skobeltsyn Institute of Nuclear Physics where they plan to work in future, to get acquainted with their future work and the main scientific problems of the different laboratories. After selection of the Group in the sixth semester, students have the possibility to attend introductory lectures in Space physics. Simultaneously the student has to select a more narrow aspect of his (her) future research work and a scientific advisor who can suggest the subject of the future diploma work and help to realize the theoretical and experimental potential of the student. Although at this stage all students are classified as either theoreticians or experimenters, but after graduation the students can work as theoreticians and experimenters. Theoreticians who have received experimental training can easily formulate the problem describing the actual situation and read experimental works to understand the accuracy of their conclusions. This give the possibility to avoid spending great efforts and time on the solution of unrealistic problems which is typical for Groups, specializing in theory. An experimenter who had the special theoretical training can understand the results of any modern theoretical work and, therefore, plan his (her) experimental investigation in the most efficient way. This becomes especially important at the preparation stage of large space research programs when adequate orbit and equipment selection can secure the success of the mission. So the combination of theoretical and experimental training gives the possibility to prepare wide-scope specialists who can conduct efficient researches.

UNIFICATION OF SPACE RESEARCH AND PLASMA PHYSICS

It the early stages of space research only a limited number of parameters was measured (mainly particle flux and then magnetic field). The development of experimental techniques gave the possibility to measure all plasma parameters including electric fields and plasma waves. Near-Earth space began to be analyzed as a huge plasma laboratory which gives the possibility to investigate processes which cannot be studied in laboratory conditions. Apart from conducting passive observations active experiments became a very popular tool of space research. This caused a restructuring of the tuition and led to the implementation at the beginning of the 1980's of a new course on the plasma physics of space processes (Antonova 1985, 1986).

THE STRUCTURE OF THE "SPACE PLASMA PHYSICS" COURSE

The great variety of space plasma parameters demands an introduction to the course, containing an overview of all the main problems of plasma physics and training of future scientists both in the magnetohydrodynamic and kinetic descriptions of plasma processes. The first part of the course is dedicated to the magnetohydrodynamic (MHD) approach. The first chapter analyzes the degree of plasma ionization degree and the structure of the Earth's ionosphere. The second chapter is dedicated to the single particle approximation. It includes the adiabatic theory of charged particle motion, adiabatic invariants and the accuracy of their conservation. The third chapter describes the properties of the particle distribution function and collision integral, and deduces the plasma transport equations and generalized Ohm's law. Three fluid, two fluid and one fluid approximations are also analyzed. It also considers the problems of magnetic field diffusion and the frozen-in condition. The anisotropic magnetohydrodynamic (Chew-Goldberger-Low approximation) approach is also discussed. The fourth chapter is dedicated to problems of magnetostatics. The magnetostatic equilibrium condition and the formalism of magnetic surfaces and magnetic flux tubes are analyzed. The magnetic stress and magnetic field pressure formalisms give the possibility to analyze magnetic pinches, current sheets and force-free magnetic fields. The virial theorem is also included. This section contains information which is not traditional for ordinary plasma courses — magnetosphere-ionosphere interaction as the main process which determines processes in the Earth's magnetosphere. The fifth chapter is dedicated to magnetic field generation in space (the dynamo problem). The sixth chapter analyzes MHD waves and MHD discontinuities including contact, tangential, Alfvénic and shock waves.

The second part of the course is dedicated to a mainly kinetic approach to describing plasmas and plasma turbulence. The first chapter "Linear oscillations and waves in space plasma" contains a detailed derivation of the dielectric permittivity tensor, describes the properties of waves in a cold magnetoactive plasma, analyzes the waves in a hot plasma and the process of their relaxation. Chapter 2 is dedicated to electromagnetic waves in space, including radio waves from the Sun and astrophysical objects, VLF and ULF waves in the magnetosphere of the Earth and waves in solar wind. Chapter 3 classifies and describes the main properties of plasma instabilities in space including fluid and current instabilities, drift and kinetic instabilities, hydrodynamic instabilities of homogeneous plasmas and parametric instabilities. Chapter 4 is dedicated to nonlinear waves and solitons in space plasmas. It includes the solutions of the Korteweg–de Vries equation and the method of Gardner–Green–Kruskal–Miura. Shock waves and solitons with an oscillatory structure are also analyzed. Chapter 5 on "Space plasma turbulence" contains sections on the quasilinear approximation, weak turbulence, strong turbulence and anomalous resistivity. Chapter 6 is dedicated to the processes of particle acceleration.

THE SELF-DEPENDENT STUDENT TRAINING

In spite of the high efficiency of the lecture course the most important part of the education is still the independent work of students. To help this process one of the most interesting subjects of the course, acceleration mechanisms in space, was selected for independent work. At the end of the spring semester the student selects a topic for his (her) report, prepares a short lecture on one acceleration mechanism. It is necessary not only describe the main features of the mechanism but also to deduce the main mathematical formulae describing the particle spectra so formed. To help the student to prepare such a report, a special textbook (Antonova *et. al.*, 1988) was published dividing all the existing mechanisms into regular and stochastic ones. Main attention was paid to seven mechanisms: betatron, induction, electrostatic field-aligned, magnetic pump, stochastic acceleration, the acceleration at shock waves and the acceleration in displaced flows. Correspondingly, the formation of the Earth's radiation belts, magnetic reconnection, the formation of double layers and a number of other processes were analyzed in detail. Therefore it is possible not only to search for the impact of the accelerating mechanism but also to learn about the structure and main properties of the space system in which it acts. The student must answer the questions of other students. Such a form of education gives the possibility to train the student not only in plasma physics but also in preparing a vast amount of material in a very compressed form and presenting it before his (her) colleagues. This will assist the student in the future preparation of his (her) diploma work and presenting the results of his (her) research at scientific conferences.

POSTGRADUATE EDUCATION

The most talented students have the possibility to continue their education as postgraduates in the Cosmic Rays and Space Physics Group. For this they must have the recommendation of the Group and pass examinations on their specialty, English language and philosophy. After receiving a postgraduate scholarship the postgraduate student has the possibility to work on a selected scientific problem for three years preparing a dissertation. This should be a high quality scientific work solving an actual scientific problem and containing new scientific results. Prior to the end of the third year of education the main results of the work must be published in scientific journals and presented at scientific conferences and scientific seminars in different institutes having space-oriented laboratories. At the end of the third year the postgraduate student must present the text of the dissertation to the Group and obtain a recommendation to present the dissertation to the Special Scientific Council. After that the commission of the Special Scientific Council reads the dissertation and either recommends it or does not recommend it for defense at a session of the Special Scientific Council. The Special Scientific Council appoints two official opponents and an opponent-organization (basically one of the institutes of the Russian Academy of Sciences) for the verification of the quality and originality of the dissertation. The postgraduate student then prepares and sends to all scientific libraries a short summary, of about 10 pages, of his (her) dissertation. Opponents and the opponent-organization prepare referee's comments on the dissertation. On the day of defense at the session of Special Scientific Council the postgraduate student presents a 20-minutes report on the main results of the dissertation, the secretary of Council reads the comments of the opponent-organization and the two opponents read their comments. Every scientist present at this open session of Council and the members of the Council have the right to discuss the quality of the dissertation and the postgraduate student's participation in the work if some results of joint works, with co-authors are presented. At the end of the session by a secret ballot the Special Scientific Council either grants or does not the degree of "candidate of physical and mathematical sciences", which is equivalent to the Ph. D. degree in western countries. The cumbersome procedure of dissertation defense ensures the high quality of the dissertations and gives the possibility to prepare scientists who can not only obtain new scientific results but also present them in a proper way. Such specialists normally continue to work in the laboratory in which they have prepared their dissertation; otherwise they can choose another place of work. In spite of the deep economic crisis in Russia, specialists with high qualifications do not generally encounter the problem of unemployment. This is connected with the enormous loss of young scientists who have chosen to work abroad or in banks and businesses where salaries are much higher than in scientific fields.

SPACE PLASMA EDUCATION PROBLEMS

One of the main problems is the limited access to computer facilities. This problem became less serious when a special computer room for student work was installed on the premises of the Skobeltsyn Institute of Nuclear Physics. Having received the possibility to work in this computer class students became proficient in modern methods of data analysis and theoretical calculations. Such a kind of education would be much more efficient if it were possible to use special educational programs. Unfortunately, at present, we have at our disposal only rather primitive and inefficient programs of such a kind. Another large problem is the material support of a student's life. Education in the Moscow State University is still free of charge, not only for citizens of Russia but also for every citizen of the former Soviet Union countries. This gives the possibility to educate in the University the most talented young people, which is especially important for space science since it demands from the scientist a very high level of talent and diligence. The crisis of the Russian economy and rebuilding of the whole life lead to very small student scholarships. Only a limited number of students have the possibility of support from the International Science Foundation. So, in spite of very cheap hostels for students and the help of parents, most students need to work. The most favourable for the student is when his (her) work is directly connected with his (her) future work. The third problem is connected with maintenance of experimental equipment. In many cases, students working on new space experiments do not have the possibility to accomplish their work since it requires the launching of a scientific satellite. Very often launches are postponed due to financial difficulties, especially in the case of large scientific missions To avoid such difficulties students and postgraduate students are usually included in scientific groups which will have the possibility to launch the satellite during the time of the student's education. If this is impossible, the students prepare the device for a future space mission and simultaneously analyze experimental data from previous missions, though this is not the best approach.

CONCLUSIONS AND DISCUSSION

In spite of the great difficulties connected with serious changes in the economic structure of Russia, the Cosmic Rays and Space Physics Group has preserved a high level of space plasma specialists training. Our former students successfully work in many scientific centers of Russia and other countries. Overcoming the economic crisis gives us the possibility to increase the curriculum time dedicated to computer training. The participation of students in the work of scientific laboratories and in the realization of large international space programs gives hope that the next generation of scientists will continue the activity started by earlier generations and greatly increase our understanding of space plasma processes.

Acknowledgments: Academician G. T. Zatsepin, Prof. I. V. Rakobolskaya, Prof. M. I. Panasyuk and Prof. B. A. Tverskoy are thanked for the attention to the work and real support.

REFERENCES

Antonova, E. E., *Plasma processes in cosmic space, Part 1,* Moscow State University, Moscow (1985).
Antonova, E. E., *Plasma processes in cosmic space, Part 2,* Moscow State University, Moscow (1986).
Antonova, E. E., M. F. Bachareva, V. N. Lomonosov, and B. A. Tverskoy, *Acceleration mechanisms in space,* Moscow State University, Moscow (1988).

Pergamon

Adv. Space Res. Vol. 20, No. 7, pp. 1387–1390, 1997

Printed in Great Britain
0273-1177/97 $17.00 + 0.00

PII: S0273–1177(97)00735–7

MOTIVATING TEACHERS AND THEIR INVOLVEMENT IN SPACE ACTIVITIES

Annette Temple

Carmarthenshire Local Education Authority, The Satellite Centre, Aberarad, Newcastle Emlyn, Carmarthenshire SA38 9DB, U.K.

ABSTRACT

It is essential that our young people are active participants in the world of space. This will only be achieved if their teachers are both well-informed and active participants themselves. A large number of teachers are currently switched on to space-related matters because of their need to use data from past and current missions. These teachers and consequently their students are no longer necessarily drawn from the traditional science disciplines. Teachers need support to maintain their motivation both to progress and to remain active in the area. The main problem to be addressed is in maintaining their involvement on a long-term basis rather than motivating their initial interest.

Background

I have been asked to address the issue of motivating teachers in space science matters and, in doing so, I will relate my experiences in working with teachers and young people in the UK schools sector during the past fifteen years. Young people are fascinated by space and over the years enlightened teachers have tried to capitalise on this interest by drawing on space applications to enhance their teaching and therefore the learning process.

It is essential that our young people have access to the latest space applications, not only to enhance their studies but also so that they can take their place in society, in the years to come, with a proper understanding of the important role that space can play in our lives. It is in everyone's interest that future generations perceive space to be relevant to their lives and, to achieve this, it is vital that the widest possible cross-section of young people, not just the most talented young scientists, have exposure to space applications in a manner relevant to them and consequently their studies.

The Current Situation

We must remember that the topic of 'Space' is alien to the majority of teachers, however many of them use space applications as tools within their subjects. This is particularly true of secondary school teachers who teach pupils in the 11-18 year range. For example, teachers of Geography now routinely use remote sensing and meteorological satellite data to enhance topics within their schemes of work. These techniques are now so widely accepted that not only is it a requirement of the syllabus, but questions are routinely set at GCSE and A-level which involve the interpretation and application of these data. These teachers are not directly interested in space science research but they do wish to have access to its products, and are becoming increasingly dependent upon its output. Space projects such as the polar platform become relevant to their lives and to those of their pupils, and therefore they are more aware of current and future research programmes and their possible impact on both their subject and, in particular, upon the teaching and learning process.

I have deliberately drawn upon an example from outside the realms of traditional science teaching, since I believe it vital that we stimulate an interest in space matters within the greatest number of young people and their teachers as is possible. This will only occur when we break out of the traditional vehicles for teaching about space.

The UK curriculum is skills-based, and relatively little emphasis is placed upon learning space-related facts, but there is a thirst developing amongst many teachers for data from space to enhance their pupils' studies. The question at the moment is whether the space and educational communities can work together in order to satisfy their increasing demands. For example, in the field of remote sensing, commercial considerations mean that many teachers can only obtain a very limited amount of data and usually it is pre-1987. These motivated teachers are therefore often disappointed, not because the data they require do not exist but simply because no special arrangements have been made for schools to have access to them. If we are to meet the expectations of these teachers, and play fair with the future generation, then urgent consideration must be given to their needs and right of access to space-related data. It is no good making a few elderly data sets available as an act of charity; young people have a need to participate more actively in space, and the immediacy of using current data sets should not be denied to them or their teachers.

In the past, schools and teachers wanted something for nothing. Now local management of schools and devolved budgets mean that monies can be allocated for purchasing data and for space-related training, but the sums are not large and commercial rates for data that apply to multi-national companies cannot be met by school budgets. It would appear sensible that consideration for access to data by schools be treated in a similar manner to that for academic institutions, where special arrangements often apply to make the data affordable. Being able to obtain even limited amounts of data would stimulate interest amongst teachers, and serve to act as an incentive to those already switched on to 'Space' to continue and expand their interest in the area. It should be remembered that motivated teachers can easily become switched off either if they fail to make the progress they expect or if they fail to receive the support/data that they were led to believe was available.

The Way Forward

Many teachers feel let down by the space community; they need data which they are not able to access and they want to involve their students in using today's space applications, but they have difficulty in finding out what is going on. Sometimes they are misled by hype about current missions and their expectations are raised, only to be dashed when they realise that it is not feasible for them to participate. With so much investment in the space segment, many rightly question why ground segment projects do not include properly planned and implemented educational programmes delivered in a way which is most helpful to ordinary teachers. So often token gestures are made towards education and teachers in particular, and often junior research team members are asked to "look after education" with no time, budget or understanding of the needs of the educational community. I would argue that very small sums of money targeted to well-researched educational projects undertaken in co-operation with informed educationalists to meet the needs of teachers would reap handsome rewards for the broader space community.

All too often large sums are spent on glossy promotional literature which is sent to schools, only to be consigned to a dusty shelf since it is not readily usable by the teacher concerned. One can always argue for larger budgets, but at the moment even the small sums currently spent are often ill-targeted, and a significant impact could be made by simply using scarce resources more effectively. Indeed, many initiatives do not require funding at all, simply consideration of the needs of teachers and how they can access data and materials in a cost-effective manner.

It is for this reason that we must all be careful to avoid hyping our activities since unrealistic expectations can be raised, and teachers who fail, in their eyes, to achieve what they believe the others have achieved will quickly become disillusioned and be unlikely to participate in any future experiment or initiative in this area. Over the years I have seen many teachers misled into undertaking projects because of the often mythological success of a few skilled individuals (whose skills have often been gained outside the realms of teaching). To pretend that "anyone can do it" when in fact some quite specific specialist skills and equipment are required is a recipe for disaster for all concerned.

Support for Teachers

Together with the need to access data and relevant up-to-date information about space, teachers need access to training to support the introduction of new techniques and learning strategies within their classrooms. Changes to the way in which teacher in-service training (INSET) is funded have given individual institutions more control over their training budgets and, whilst certain restrictions apply, in practice individual teachers are able to identify their training needs, giving far more flexibility than national programmes have allowed for in the past.

However, there are few educational trainers who provide space-related training and this area needs expanding if teachers' needs are to be adequately met. In addition to this, teachers require good quality support materials which are readily usable in the classroom to enhance the teaching and learning process. These materials should not only support specific courses but allow for participation by teachers and pupils with different levels of expertise and abilities respectively.

One illustration of this is in the involvement of our youngest pupils, aged 4-11. Their young minds are exceptionally receptive to the wonders and challenge of working in the space environment. They and their teachers need special thought and attention when support materials are being originated. Most teachers at Primary level teach in a thematic, multi-disciplinary manner, giving enormous scope for the introduction of 'Space' as a main or sub-theme. Primary school teachers are already including space studies in themes such as Weather, Landscape, Satellites, Travel, etc.. However, there is an acute shortage of materials with the correct approach and language levels appropriate to this age group. Much work remains to be done in this area and the current demand for materials remains largely unmet. The origination of support materials is clearly a task to be undertaken by educationalists but they, in turn, will require a high level of co-operation, inspiration and help from those engaged in current missions and research programmes.

Many schools in both the secondary and primary sectors are now equipped to access the Internet. It remains to be seen whether access to this technology will allow teachers to access more of the background material and/or data which they require. There is certainly a challenge for us all to try to ensure that teachers needs are considered when uploading information onto web sites. Perhaps specialist interest groups can be persuaded to contribute to a space education web site which will serve to break down the isolation felt by many teachers working in the area and to stimulate and encourage others.

Video and personal conferencing via British Telecom's Integrated Services Digital Network (ISDN) is becoming more common in secondary schools. The opportunities for interchange of ideas, mentoring and remote support from specialists are still being assessed, but it is an area which we cannot afford to ignore and which could offer a cost-effective way of providing much needed links between schools and those working in the space industry.

Building Bridges

National Space Agencies have, in my view, a particular responsibility to act as a bridge between the two communities and to facilitate joint initiatives to the benefit of all involved. I am not advocating a bureaucratic approach to the matter, simply that effort should be expended at this level to support interested parties within both the space and educational communities to come together for their mutual advantage. The benefits for all concerned are potentially enormous. In addition, conferences have a significant role to play, particularly if we can use the event to forge new links between educationalists and those with an interest in education within the space community. If we work together to better understand each others needs, problems, worries and concerns, I am sure that much can be achieved using existing resources. The problem, as I see it, is therefore not concerned with motivating teachers in the first instance but in supporting those who are already switched on to 'Space' so that they continue working in the area and don't become disillusioned and drop out. Success breeds success. Only by identifying and disseminating good practice can we hope to achieve our objective of ensuring that all teachers and their pupils are not only aware of the impact of space on our society but are active participants in its progress.

Pergamon

Adv. Space Res. Vol. 20, No. 7, pp. 1391–1396, 1997

Printed in Great Britain
0273-1177/97 $17.00 + 0.00

PII: S0273–1177(97)00736–9

GLOBAL CHANGE SCIENCE: ISSUES RELATED TO EDUCATION AND TRAINING

S. I. Rasool* and J.-L. Fellous**

**JPL/UNH (U.S.A.), Current address: 52, Bld Sébastopol, 75003 Paris, France*
***MEDIAS-FRANCE, CNES-BPi 2102, 18, Avenue Edouard-Belin, 31401 Toulouse Cedex 04, France*

ABSTRACT

The very latest assessment of the Intergovernmental Panel on Climate Change (IPCC) is that the global climate will continue to change in the next century and that "a discernible human influence on climate can now be identified". If one is to accept this conclusion, scientific discussion should now focus on what kind of change is to be expected for different regions of the world. What will be the impact of these climatic changes on ecosystems, agriculture, commerce and on the social systems ? How can one mitigate or adapt to these changes ? In addition, this discussion needs now to be carried from research institutions to schools and colleges so that the next generation is aware of not only the scenarios in which it will be living but the options which can be available to adapt to the changing habitats. A good beginning in this regards has been made by the Education in Global Change Project of ICSU.

THE SECOND IPCC SCIENTIFIC ASSESSMENT OF CLIMATE CHANGE

In the last several decades atmospheric concentrations of greenhouse gases have continued to increase. These increases constitute a significant enhancement of the greenhouse effect, leading to complex climate change. In fact a second scientific assessment of climate change carried out by the Intergovernmental Panel on Climate Change (1995) states that the global mean temperature increase over the last century "*is unlikely to be entirely due to natural causes*", and that "*a pattern of climate response to human activities is identifiable in climate records*". Also they assert that "the global climate will continue to change into the next century and beyond" (Figure 1).

On the other hand aerosols in the lower atmosphere (resulting from combustion of fossil fuels, biomass burning, etc.) exert locally a cooling effect on climate (global average 0.5 Wm^{-2}) but, in contrast to most greenhouse gases aerosols are very short-lived; hence their radiative forcing adjusts rapidly to increases or decreases in emissions. It has also been found that in 1994-95 over Antarctica ozone soundings showed near complete destruction at altitudes between 14 and 21 km (WMO, 1996). The area of the ozone hole (over 22 million km^2) was similar to the 1993 record, and the duration was longer than ever previously. Finally the relatively low global mean surface temperature observed in 1992-93 was due, at least partly, to a large, transient perturbation induced by the Mount Pinatubo eruption in mid-1991. The temperatures have largely recovered since.

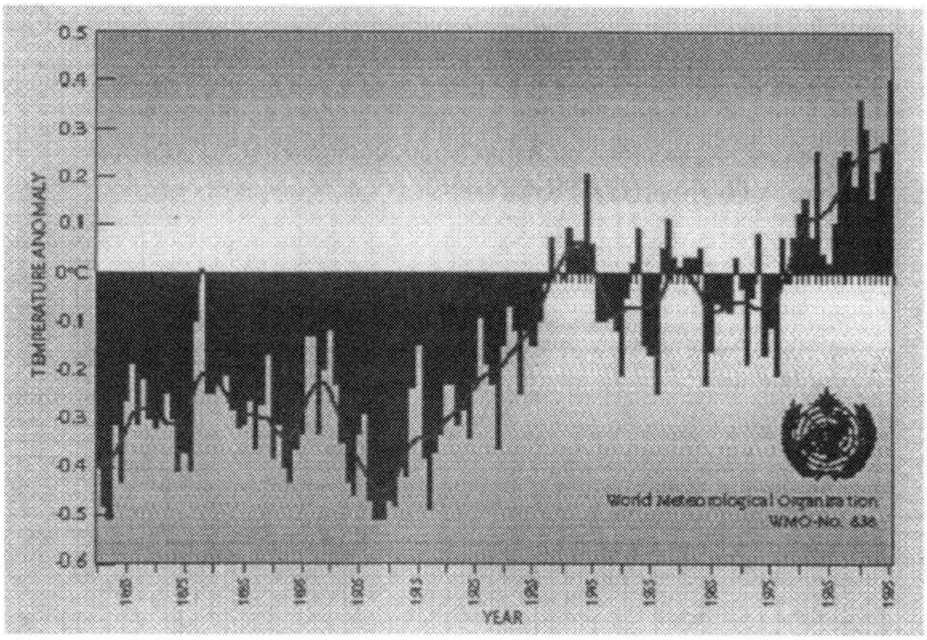

Fig. 1. 1995, globally, the warmest year on record *(after WMO, 1996)*

GLOBAL CHANGE AND REGIONAL IMPACTS

Global change models have continued to improve in the last decade. They are now coming close to reproducing the changes which have occurred over the globe during the last 40 years. With this renewed confidence model projections of future climate change are becoming somewhat more credible and provide useful information on future climate change. However, confidence in regional climate changes predicted by models remains low. The recent IPCC assessment of future climatic changes concludes that "global mean surface temperature will increase at a rate between 0.12 and 0.33°C per decade" (i.e., an increase of about 2°C by 2100 relative to 1990). It also states that regional changes may differ substantially from the global mean value: warming is greater over land than oceans, maximum warming should be expected in high northern latitudes in winter.

For sea level changes they predict a rise by between 0.2 and 1.0 m by the year 2100, as a result of thermal expansion of the oceans and melting of glaciers and ice sheets. It is also stated that sea level will continue to rise for several centuries thereafter, even beyond the stabilisation of global mean temperature.

Many uncertainties in predicting climate change of course remain. These include the response of clouds to global warming, the role of ocean in modulating future climate change, the role of changes in vegetative land cover, the magnitude of natural climate variability, the past history of radiative forcing by greenhouse gases and sulphate aerosols. Despite these uncertainties in the prediction of the future climates it is instructive to begin reflecting on the range of possible impacts of future changes on habitability of the Earth. This is because most systems (natural ecosystems, food production, water resources, etc.) are sensitive to climate change (Figure 2). For example 1-3.5°C is equivalent to a 150-550 km poleward shift or to a 150-550 m altitude shift. Therefore the geographic distribution and composition of many ecosystems will shift as individual species respond to climate change. The IPCC assessment also implies that climate change will lead to intensification of the global hydrological cycle and can have major impact on water resources. Moreover, high latitude regions may experience increased runoff, due to increased precipitation, whereas at lower latitudes runoff may decrease due to increased evapotranspiration and decreased precipitation, and water availability will fall under 1,000 m^3 per person per year in many regions. Changes in water availability is bound to impact patterns of crop production (Table 1), leading to large differences at local and regional scales.

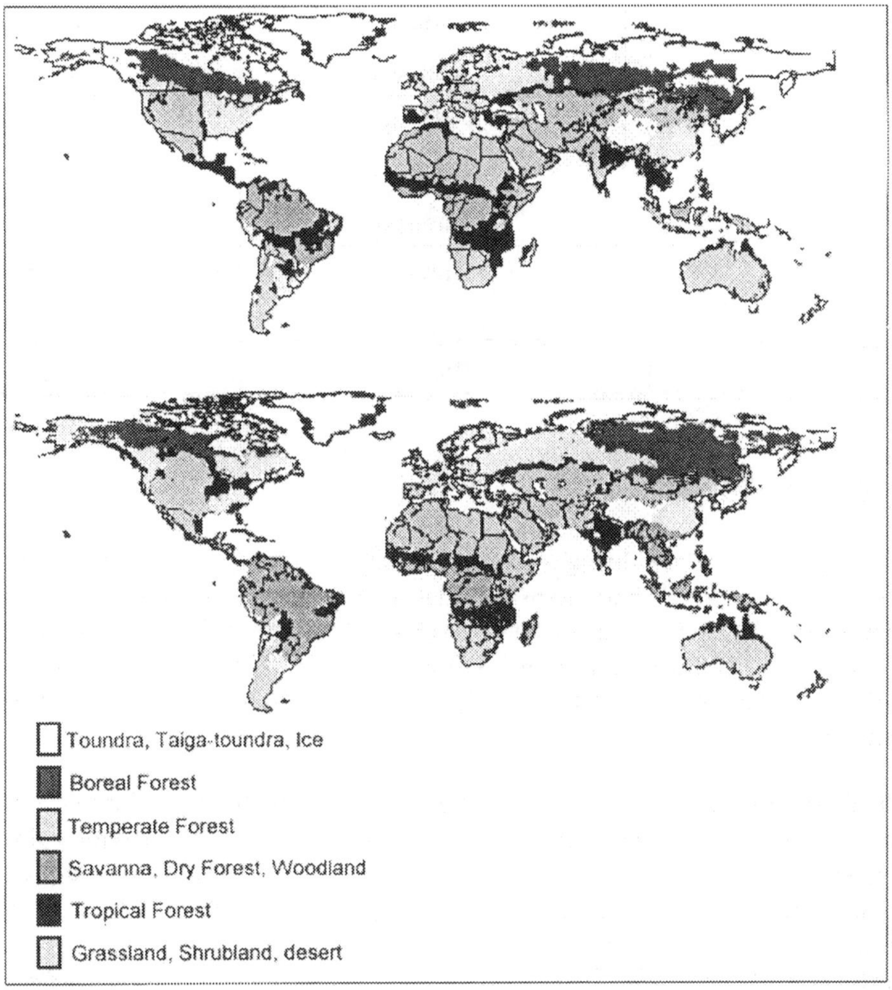

Fig. 2. Potential distribution of major biomes (top) current climate (bottom) $2xCO_2$
(after IPCC Report Summary, 1996)

LIVING WITH GLOBAL CHANGE

Although changes so far have been slow and subtle, the future projections are large and sometimes drastic. It is clear that next generations will have to learn "*how to live with global change*". Fortunately our "*Watchful Eyes*", i.e. the space-based measurements of the Earth, are now global and have become accurate enough that they can be used to monitor precisely the evolution of the planet on regional scales. This will certainly help us to understand the change, adapt to it and may be mitigate. The global community should now get involved into hands-on assessment of how the region of interest is evolving with time, compiling data sets for diagnosing the observed change with cause and effect argumentation, and, developing adaptation strategies and mitigation options.

Table 1. Expected impacts of doubling CO_2 on selected crop *(after IPCC Report Summary, 1996)*

Region	Crop	Yield impact (%)
Latin America	Maize	-61 to increase
	Wheat	-50 to -5
	Soybean	-10 to +40
Africa	Maize	-65 to +5
	Millet	--79 to -63
	Biomass	decrease
South Asia	Rice	-22 to +28
	Maize	-65 to -10
China	Rice	-78 to +28

NEW DATA SETS AVAILABLE

Data sets for this purpose are now being made available at cost of reproduction, in order for College and University students to develop their own activities and projects, e.g., the NASA/NOAA Pathfinder products (Figure 3), or the "Africa and global change" diskette/CD-ROM project (NOAA/NGDC, MEDIAS-FRANCE). IGBP/DIS has triggered the development of new global data sets, such as global land cover, global fire products, global soil maps, global topography (Figure 4), global NPP-GPPDI, global wetlands, etc.

College and University students may play a significant role in combining satellite-provided data sets with local/regional *in situ* data to derive data *products* of regional practical interest. Specific training is required in order for students to accomplish this role, and training for teachers is a prerequisite. For this purpose several training initiatives are currently underway which can serve as good examples to follow in other parts of the world. Among these are:

- an excellent text published by ICSU's Education in Global Change Project, "Science for understanding tomorrow's world: Global Change" - student activities and teachers guide, ICSU (1994)
- the International Summer University, Toulouse, 15-27, July, 1996 (CNES, French Ministry of Education) : "Satellite contribution to the observation, understanding and management of the Earth environment at global, regional and local scale",
- the International MEDIAS/START/DIS Workshop, West Africa, 1997 : "Data management for global change in Africa",
- the International MEDIAS Schools, held in Niamey, Niger (1993): Subtropical climates and their evolution - from space-borne observation to modelling, and in Nairobi, Kenya (1995): Africa and global change, or in preparation (Cape Town, South Africa, 1997 : Satellite oceanography - A tool for global change research projects in Africa),
- the GAIM Africa Workshop, Nairobi, Kenya, February 1997, with focus on modelling and data analysis of relevance for Africa.

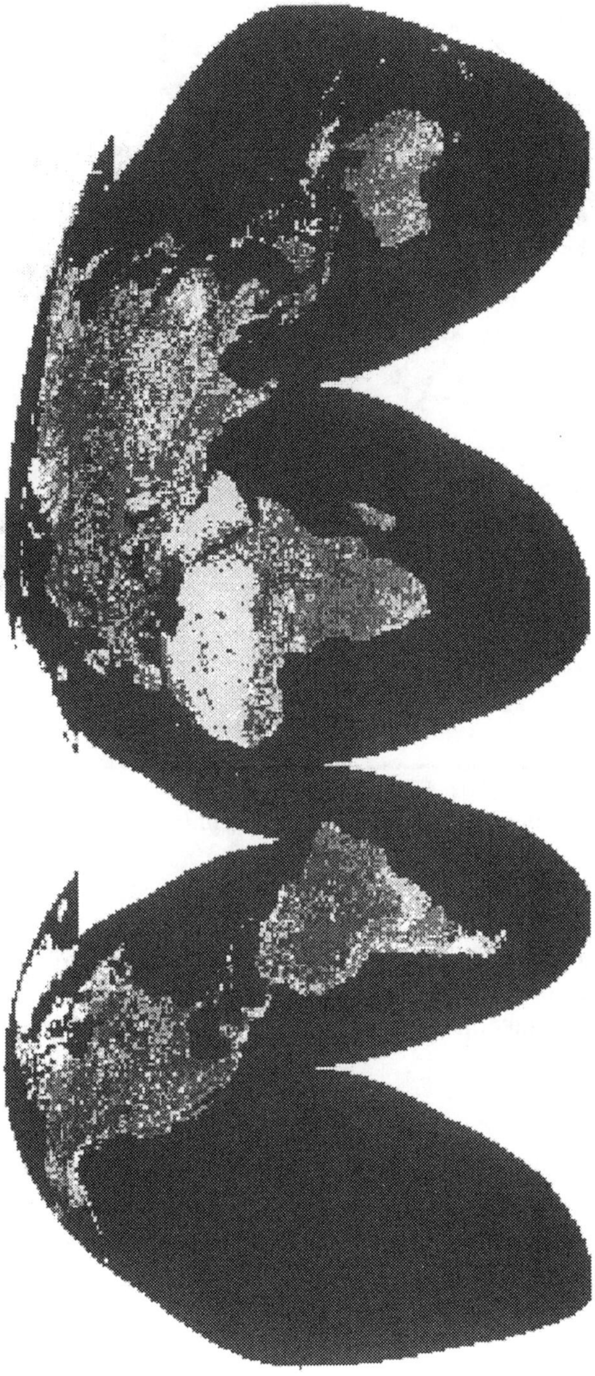

Fig. 3. AVHRR Global 1km 10-day composites over 12 months now available free of charge at:
http: /edcwww.cr.usgs.gov/landdaac/1KM

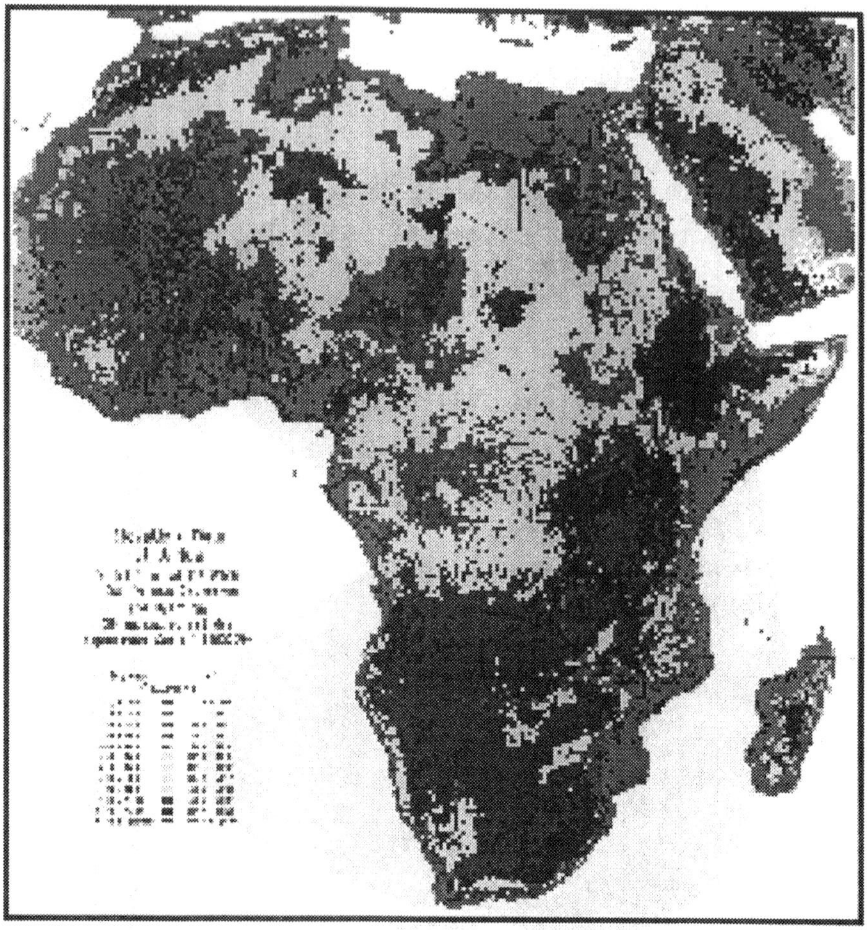

Fig. 4. Digital elevation model of Africa *(after NASA/GSFC)*

CONCLUSION

Accurate satellite data are ideal to document precisely how the world is changing from day to day, month to month and year after year. The sooner we learn how to handle them and use them to understand how our region is responding to the changing climate, the better we will be able to "manage" the change.

REFERENCES

IPCC 2nd Assessment, Climate Change 1995, Reports of IPCC Working Groups I, II and III (1995).
WMO Statement on the status of the global climate in 1995, Report WMO-N°838 (1996).
Science for understanding tomorrow's world: Global Change, Ed. David Waddington, ICSU Press (1994)

Pergamon

Adv. Space Res. Vol. 20, No. 7, pp. 1397–1399, 1997

Printed in Great Britain
0273-1177/97 $17.00 + 0.00

PII: S0273–1177(97)00737–0

EDUCATION PROGRAMME ON AEROSPACE AND ENVIRONMENTAL MEDICINE FOR MEDICAL FACULTY OF LOMONOSOV MOSCOW STATE UNIVERSITY

A. I. Grigoriev, L. B. Buravkova, V. A. Loginov and O. L. Vinogradova

Aerospace and Environmental Medicine Chair, Space Biomedical Centre for Training and Research, Moscow, Russia

ABSTRACT

The problems and approaches to organisation of the education process in the field of aerospace and environmental medicine for medical students are discussed. Original education programs are developed on the basis of Russian experience in space biology and physiology, environmental medicine, aerospace medicine and medical support during spaceflight. The main goals of these programs are to acquaint students with: interaction of living organisms with natural and artificial surroundings, including space flight conditions; the physiological reactions on extreme environmental factors; basic mechanisms of human adaptation to space flight and particularly to microgravity; the current research in space medicine and new telecommunication technologies. All programs are formed in accordance with contemporary progress in life sciences and revealed a result of the interdisciplinary approach to education process.

INTRODUCTION

In Russia high medical education implies intensive practice in traditional clinical disciplines and comparatively short courses on such branches of life sciences as theoretical and molecular biology, modern biochemistry and genetics, mathematical and physico-chemical methods in medicine, application of new methods to environmental medicine. To improve the situation the Faculty of Basic Medicine was formed in Lomonosov Moscow State University in 1992. It is planned that after seven (6+1) years' study at University students will be able to use acquired academic knowledge in hospitals. But one of the major problems confronting a fully educated physician is how to apply the basic knowledge, obtained at University, into clinical practice.

To solve this problem, the Space Biomedical Centre for Training and Research (SBCTR) was organised under the supervision of Russian Ministry of Science and Technical Policy, Lomonosov Moscow State University and Institute of Biomedical Problems (Moscow). One of the main parts of this Centre is the Aerospace and Environmental Medicine Chair (ASEMC).

Several training courses are planned to be taught to the senior students and postgraduates of Faculty of Basic Medicine by ASEMC: Space Biology and Medicine, Environmental Medicine, Telemedicine, etc. Some of the lessons will be focused on the results of space life sciences experiments that were designed and carried out by Russian and American investigators.

Each of the courses has an original education program, developed specially for Lomonosov Moscow State University students on the basis of the research experience obtained by specialists from Institute of Biomedical Problems. The programs are accomplished by lectures, demonstrations, seminars, small group discussions, library research and individual reports.

GENERAL PRINCIPLES OF THE EDUCATION PROGRAMS

We do not aim at providing the background necessary for the proper practice in different medical disciplines. The main purpose of all the programs in the field of aerospace and environmental medicine is to make students familiar with the physiological interrelations of human beings with their surroundings through changing conditions of existence. Also during the training courses the principles of biomedical correction of pathophysiological reactions of organisms to different extreme factors will be widely performed.

Among the large spectrum of general tasks which are put before students while studying at ASEMC there are:

a) a thorough study of the physiological peculiarities of human organism interrelations with the natural and artificial environment (long-time expedition, spaceflight, etc.);
b) the study of the main adaptation mechanisms at cellular and subcellular levels;
c) individual training in life sciences modelling and investigation methods, widely used in gravitational physiology, aerospace and environmental medicine;
d) introduction to medical support special professional categories (cosmonauts, astronauts, pilots, divers etc.) to obtain the knowledge for and principles of proper application of modern life sciences results into clinical practice;
e) to get grounding in International Medical Programs and to be able to use contemporary experience of telecommunication (Telemedicine) for a physician's everyday purposes.

According to these principles we developed programs which can be used together as well as separately in the educational process.

Extreme Environmental Factors in Physiology and Medicine

This program contains a general classification of extreme influences, including geographic factors. The physiological reactions of changes in barometric pressure, temperature, ionized and non-ionized radiation, gas and chemical composition of the surroundings are considered in detail. Much attention is paid to the role of biological rhythms and adaptation. The methods of medical correction of pathology caused by extreme factors are studied.

Space Biology and Medicine

The lessons begin with a theoretical course dealing with the role of gravitation in the development of living organisms. Then the ethiology and pathogenesis of changes in tissues, organs and systems (heart and renal function, muscle atrophy, water-electrolyte balance, neuroendocrinal regulation, etc.) during short- and long-term spaceflights are studied in a comparative manner. The third part of this program contains a practical course of crew medical support, cosmonauts training and biomedical control during spaceflight.

Life Support Systems

This is a mainly practical course which includes: review and classification of modern life support systems, the ecological regulation of artificial environment, the principles of the long-term interaction of organisms with their physical surroundings. Also the medical aspects of extravehicular activity and the construction of different space suits are studied.

Telemedicine

The main principles of medical consulting by means of a telecommunication network are studied. It is also planned to study quality, limitation and effectiveness of medical aid with the help of computer multimedia stations. Telemedicine should not be limited to distant consultation, either on-line or off-line, and transferring remotely acquired biomedical data. Another strong point of the telemedicine course is information support for physicians.

DISCUSSION

All education programs are formed in accordance with the current progress in life sciences. To a great extent the programs are flexible and interchangeable. Focus (blocks of themes) is used in the educational process. Each block contains approximately 70% of new information and 30% of information which is corresponding to the main items of the program (the holographic principle). The practical work takes a half of the education time and is organised with the Institute of Biomedical Problems (Moscow).

Increasing contacts between Lomonosov Moscow State University and leading biomedical establishments in the field of aerospace and environmental medicine are of great mutual benefit. The appearance of well prepared Russian physicians, familiar with modern trends and methods of research, and understanding specific aspects of clinical practice, will be the result of such co-operation.
The benefit of the program are illustrated in Table 1.

The postgraduates of the
"AEROSPACE AND ENVIRONMENTAL MEDICINE CHAIR"
will be able to use

the results of Space Life Sciences experiments in clinical practice	the principal biomedical correction of pathophysiological reactions to extreme conditions	the application of space methods and equipment in medical consultations

Pergamon

Adv. Space Res. Vol. 20, No. 7, pp. 1401–1404, 1997

Printed in Great Britain
0273-1177/97 $17.00 + 0.00

PII: S0273–1177(97)00738–2

EDUCATION IN ASTRONOMY AND ASTROPHYSICS: THE INDIAN EXPERIENCE

J. V. Narlikar and N. C. Rana*

Inter-University Centre for Astronomy and Astrophysics (IUCAA), Post Bag 4, Ganeshkhind, Pune 4110078, India
**Deceased, August 22, 1996*

ABSTRACT

In the context of a vast region with one-sixth of the global population, the Indian subcontinent has a very rich tradition of teaching of astronomy as a science. The so-called modern astronomy has mostly been developed over the past two centuries whereas the subject of astrophysics is barely a hundred years old. This work reviews briefly the current efforts in India in propagating these subjects not only at the research level but all the way down to the level of school children. With the creation of IUCAA, these efforts have been systematized and the response to them has been very rewarding.

INTRODUCTION

India has a very long and enduring tradition for teaching and research in astronomy. The recent review by Kochhar and Narlikar (1995) summarizes the historical developments. In the Vedic times, dating back more than two and a half millennia, there were works like the *Vedanga Jyotisha* which carried astronomical ideas and speculations. During the golden period of Indian astronomy ranging from Aryabhata (5th Century) to Bhaskara II (12th Century) astronomy had advanced from calendar making to observations of celestial bodies with special instruments called *Yantras* and usage of trigonometric and algebraic manipulations. It is worth noting that Aryabhata had clearly stated that in his view the Earth spins about an axis against the background of fixed stars ... an idea that ran counter to the then prevalent geocentric theory.

Then there was a lull for a few centuries and a revival of astronomy in India came with the colonial era about two hundred years ago, with inputs from the French and the British. Telescopic astronomy came with the Europeans and there were several expeditions from Europe to observe special events like occultations, conjunctions, eclipses, etc. The discovery of the element helium was a by-product of one such expedition in the last century. The usage of spectroscopy in astronomy was a precursor to astrophysics. Here India can claim to have taken the initiative through the work of Meghnad Saha on the ionization equation (de Vorkin, 1993).

The major achievements in astrophysics came through the works of Saha, S.Chandrasekhar and D.S.Kothari in the twenties and the thirties. This was the legacy on which the post independence (post-1947) developments in astronomy and astrophysics (A&A) have been founded. The current scenario in these fields in India will be described next.

THE A&A INFRASTRUCTURE IN INDIA

The present infrastructure in astronomy and astrophysics in the country can be described as a combination of three sectors:

- The research institutes and observatories
- Universities and colleges
- The amateur astronomy movement

We briefly outline these three areas.

The research institutes and observatories:

These include autonomous institutes set up by the Government of India through its Departments of Space, Atomic Energy and Science and Technology. These institutes have been partially or wholly devoted to research in A&A. They train research students for Ph.D. degrees in A&A while most of them also run a few observational facilities. Table 1 gives a brief summary of these institutions:

Table 1. Indian A & A facilities (de Vorkin, 1993)

Institution	Partially (P) or Wholly (W) devoted to A&A	Observing facilities
Tata Institute of Fundamental Research, Mumbai and Pune	P	Balloon Facility Radio Telescopes Gamma ray Detectors
Physical Research Laboratory, Ahmedabad	P	Solar Observatory Radio Array
Indian Institute of astrophysics, Bangalore	W	Optical Telescopes
Raman Research Institute	P	Radio Array, MM-Dish
Uttar Pradesh State Observatory, Naini Tal*	W	Optical Telescopes

* This observatory is managed by the State of Uttar Pradesh

The universities and colleges:

In the pre-independence India the subject of spherical astronomy was taught in the mathematical departments of several universities. This continues even today. However, with the emergence of astrophysics as an important branch of physics several universities now include it as an optional part of the Master's course in physics. Today more than twenty universities have A&A in some form or other as part of their M.Sc. programme in either mathematics or physics. A few colleges affiliated to the universities also run such courses.

There are two universities, Osmania University, Hyderabad, and Punjabi University, Patiala, which have departments wholly devoted to astronomy. These universities have played a major role in training astronomers and astrophysicists, many of whom are active on the Indian scene today. The Osmania University also runs an optical observatory with a 1.2 metre telescope at Japal-Rangapore within an hour's drive of the university.

However, over the years the university-sector was getting neglected in the overall growth and development of A&A. To initiate a revival, the University Grants Commission created in 1988 a new institution at Pune called the Inter-University Centre for Astronomy and Astrophysics (IUCAA in brief) with a mandate to nucleate and promote the teaching, research and development of A&A in the university sector. We outline IUCAA's activities in the following section.

The amateur astronomy movement:

While the first two streams broadly cater to the professional astronomy community, which today is around 300-350 (including graduate students), there has traditionally been a large amateur interest in A&A in India (Kochhar & Narlikar, 1995).The amateurs have either worked individually with optical telescopes in the 8-15 cm class, or in voluntary organizations or clubs.

Till recently, this sector was diffuse and unorganized. Largely through an initiative taken by the IUCAA in 1991, the various amateur organizations have been trying to come together under the joint umbrella of the Confederation of Indian Amateur Astronomers. The Federation organizes an annual meet of the amateur community in which a few professionals also participate. The amateurs play a key role in the Indian society for dissemination of astronomical information and on special astronomical events like solar eclipses, cometary visits and so on.

IUCAA's PEDAGOGICAL ACTIVITIES

We now outline the broad spectrum of activities of IUCAA which are aimed at growing human resources in A&A at all levels. Table 2 summarizes these. We should mention that the IUCAA has freely drawn the resource persons for its pedagogical activities from the research institutes in India and scientists from abroad as well as from universities and the amateur community.

Table 2. Pedagogical Activities of IUCAA

Level of participants	Activity
Post-doctoral, research students	Advanced, state of the art workshops
Introducing new research areas to graduate students	Mini-schools
Preparing graduate students for research in A&A	Inter-university graduate schools
University teachers in A&A	Refresher courses
Undergraduates	Introductory schools in A&A
High School students	Summer projects and also monthly lectures in A&A
School teachers and amateurs	Workshops for training the basics of astronomy, including hands-on training in making sky-globes and small telescopes

The IUCAA has excellent facilities for holding such meetings on its campus and several are held there round the year. However, in keeping with the desire to bring A&A closer to the universities and colleges, many of these meetings are held in their campuses also.

Apart from these institutional activities, several Indian astronomers (including the authors of this paper) also lecture individually on various public forums. The enthusiasm of the general public towards astronomy appears boundless and has not been adequately tapped so far.

CONCLUSION

India is an unusual country in the sense that it presents the sharpest contrast between a highly developed society and a highly primitive one coexisting side by side. The contrast is not merely one of economic development. The lack of scientific temper and adherence to superstitions like astrology are found in the highest strata of society. This is where astronomy education has a major contribution to make.

In this context, there is a strong need to enhance the astronomy part in the science education curricula of Indian schools. Through astronomy one learns to appreciate the vast scope for the application of the laws of science, one realizes the smallness of the Earth in the overall scheme of the cosmos and, last but not the least, one is forced to appreciate the need to make our planet safe and habitable: for as yet it is the only known habitat for life.

REFERENCES

deVorkin, D.H.: Saha's influence in the west in Meghnad Saha Birth Centenary Commemoration volume, Ed.S.B.Karmohapatro, Saha Institute Publication (1993)

Kochhar R.K. and J.V.Narlikar: Astronomy in India: A perspective, Indian National Science Academy, New Delhi (1995)

Pergamon

Adv. Space Res. Vol. 20, No. 7, pp. 1405–1410, 1997

Printed in Great Britain
0273-1177/97 $17.00 + 0.00

PII: S0273–1177(97)00739–4

THE PROSPECTS OF ENHANCED SPACE SCIENCE TRAINING IN KENYA

J. O. Aseno* and J. D. Obel**

**Department of Surveying, University of Nairobi, P.O. Box 30197, Nairobi, Kenya*
***Survey of Kenya, P.O. Box 30046, Nairobi, Kenya*

ABSTRACT

To a limited extent, space exploration has been conducted in Kenya for almost the last two decades through a joint project (San Marco Project) between the Government of Kenya and the Government of Italy. Other space science activities in the country include remote sensing, space communications, meteorology and the use of navigation and positioning satellite systems.

To sustain space science activities in Kenya will require specialized training in the various disciplines of space sciences. Currently, there are no well coordinated training programmes in the country. Consequently, there is an urgent need for a well planned and a well coordinated space science training programme. This could be achieved through international co-operation and joint ventures between Kenya and space science institutions/organizations worldwide.

The paper justifies the need for training in space science in Kenya and discusses socio-economic as well as environmental gains which would be realized due to increased space science activities arising from such training. Some of these gains would include participation in the launching and tracking, and control of satellite, managing and running a space centre or satellite launching and tracking station, decoding and synthesizing data from satellites and disseminating such data for public and scientific uses.

The paper further offers suggestions on how the training requirements cited above could be achieved. It also highlights the level of expertise in space science disciplines and provides specific recommendations on the types of personnel that need to be trained. In addition, various forms and levels of training required to strengthen the role of space science in socio-economic development in Kenya, are discussed.

INTRODUCTION

The application of space science and technology in Kenya started in 1964 with the signing of an agreement between the Government of Kenya and the Italian Space Commission on a joint venture space science project. The project was referred to as the "San Marco" launching range and was based in the Kenyan town of Malindi. Arising from this, the first satellite was launched in 1967 and, over the years, a total of twelve satellites and sounding rockets have been launched from San Marco.

In the early 1970's, Kenya started her involvement in geodetic positioning using Doppler Satellites, through an experiment which was conducted jointly by the British and American military survey teams. Following the success of the experiment, Survey of Kenya acquired some Doppler satellite receivers and conducted limited Doppler satellite positioning surveys. This effort was further supplemented by the African Doppler Surveys (ADOS) project which was undertaken between the period 1981 to 1986. The project was sponsored by the International Association of Geodesy (IAG) and coordinated by the Regional Centre for Services in Surveying, Mapping and Remote Sensing (RCSSMRS), in Nairobi. It was meant to provide a common geodetic control framework for the African Continent.

At present, the Global Positioning System (GPS) has taken over from Doppler Satellite Positioning in Kenya. However,

GPS has not had much impact due to prohibitive cost of receivers and computing facilities coupled with the lack of adequately trained staff.

In terms of space communication, the Longonot I Earth station for satellite radio communications was commissioned in 1970 as a regional telecommunication centre serving three East African countries: Kenya, Uganda and Tanzania. Later, Longonot II was added to provide a global communication network. Currently a new station is under construction at Kericho in the Rift Valley province.

Pioneer work in remote sensing was also conducted for the first time in Kenya in 1970's. This, in later years, culminated in the establishment of a Department of Resource Surveys and Remote Sensing (DRSRS), which together with RCSSMRS and Survey of Kenya, coordinate and manage the application of remote sensing technology in resource surveys.

Meteorological data from weather satellites have also been utilized in Kenya since the mid 1970's. This has been through Automatic Picture Transmission (APT) equipment which receives the Earth's cloud images from the polar orbiting American National Ocean and Atmospheric Administration (NOAA) Satellite. Over the years, due to the installation of the High Resolution Picture Transmission (HRPT) receiver, the quality of the received image from NOAA satellite has appreciably improved. In addition, the Primary Data User Station (PDUS) equipment was installed for METEOSAT satellite reception, in order to provide cloud imagery and other environment data at 30 minute intervals.

Despite the early start in space science activities in Kenya, there has been no major impact on socio-economic development. This has been due largely to a perennial lack of adequate personnel training at the various levels of production. This situation has resulted in a serious underutilization of the potential of the technology for the betterment of the country's economy and socio-political aspirations.

THE ROLE OF SPACE SCIENCE IN SOCIO-ECONOMIC DEVELOPMENT

Space technology has the capacity to play a major role in surveying and mapping since the use of the technology would enhance the process of data gathering, storage and retrieval. This, in turn, would lead to improved efficiency in the planning and execution of projects since the space techniques presently provide relatively better efficiencies and accuracies than conventional methods in terms of determination of positions of points on the surface of the Earth, preparation of photographic images and maps, production of special imagery applicable in mapping, land-use planning and environmental monitoring and protection. These techniques include Global Positioning System, Photogrammetry, Remote Sensing, etc.

To guarantee peace and security between neighbouring countries, it is important that the common boundaries are defined, demarcated, surveyed and mapped. In this regard, space technology has the potential to achieve the desired results rapidly. In addition, space technology may be used to monitor the crustal deformation patterns of the Kenyan Rift Valley in order to assist in disaster mitigation efforts when earthquakes or volcanic eruptions occur. Such efforts would help to ensure the safety of human beings and those of economic investments within the rift valley when natural disasters occur.

The extensive environmental degradation resulting from global climatic and tectonic changes and unsustainable use of the natural resources, has adversely affected the socio-economic development of many countries, especially in Africa. Space technology, particularly remote sensing, makes it possible to collect and analyze information about resources and the environment over large study areas.

The remote sensing technology is applicable in the following areas, among others: land-use planning and mineral exploration, vegetation and agriculture, water and coastal zone management, marine resource management, environmental monitoring and protection, natural disaster reduction and economic planning.

The major role of space communication is to provide international communication services, which are indispensable aspects of socio-economic activities in any country. Satellite communication is applicable in telecommunications, broadcasting, meteorological data collection, space research, navigation and resource exploration. These are vital parameters in the socio-economic development of a nation. In addition, satellite communications benefit mankind in the area of distant education exchange programmes and in transmission of scientific data for research in medical services, agriculture, military surveillance, etc.

Furthermore, full implementation of space technology in the various sections of a Meteorological Department would result in a National Meteorological and Hydrological Services (NMHS), providing oceanographic and other environmental data and services. These would include daily weather forecasts, medium and long range weather forecasts, aiding national economic planning and management programmes. These attempts would also contribute to the worldwide efforts to monitor, understand and predict weather and climate for the implementation of reliable and sustainable development projects.

CURRENT STATUS

At the moment, there is no space technology being applied adequately in the field of surveying in Kenya. Limited application of space technology was attempted using the Doppler Satellite technique one and a half decades ago. In the recent past, GPS came into the market worldwide, but its application in Kenya has been minimal due to the lack of personnel training and limited financial capability. Consequently, no nationally based GPS operation has been undertaken. This is quite unfortunate considering that, out of the six Kenyan international boundaries, only four (with Uganda, Ethiopia, Somalia and Tanzania) have been surveyed. The other two (with Sudan, and in the Indian Ocean) are yet to be demarcated, surveyed and mapped.

Furthermore, Kenya has the 12-nautical mile Territorial sea in the Indian Ocean, measured seaward from the baseline along the shore, in accordance with the UN law of the sea convention of 1982. The territorial sea portion has an area of 14,300 square kilometres and therefore increases Kenya's total area of 587,900 square kilometres (576,700 of land and 11,200 of inland waters) by 2.4 percent to 602,200 square kilometres. In addition, Kenya has the 200-nautical mile Exclusive Economic Zone (EEZ) measured seaward from the said baseline. In accordance with the said convention, Kenya has exclusive economic rights to the EEZ, which covers an area of 143,100 square kilometres. It is regrettable that it has not been possible to use space technology to survey and map those areas for economic gains and international peace and security.

However, since 1976, various government and non governmental organizations have been involved in space technology in the interdisciplinary areas of resource and environmental management. This has led to limited application of technology in the following areas: (i) mapping and assessment of the state of natural resources, (ii) making an inventory of renewable and non-renewable resources (iii) forecasting crop yields and providing an early warning system for drought, food security and desert locust invasion (iv) detecting crop diseases and pest control, (v) assessing and mapping biomass, forest cover mapping and detecting rate of depletion, and preparing forestry master plans, (vi) producing land-use maps (vii) producing exploratory soil maps, and (viii) preparing wildlife habitat maps. Application of space technology needs to be intensified in these areas in order to increase economic gains and social benefits.

In space communication, Kenya is covered by the INTELSAT international public communications services, based on a global communication satellite which was launched in 1965. The use of satellite communication in Kenya started in 1970 with the commissioning of Longonot I INTELSAT satellite Earth station. This was later followed by the commissioning of Longonot II in 1980. To date, the two Earth stations provide mixed analogue and digital international public communications services for telephony, telegraphy, telex, facsimile, data transmission and occasional television and radio programmes.

The meteorological service in Kenya has four main components which include (i) a network of observing stations, (ii) a system of communications, (iii) a national meteorological centre, and (iv) a regional meteorological centre. Meteorological data obtained from the four components are analyzed at the national meteorological centre in Nairobi. The findings are utilized in daily weather forecasts for public awareness, aviation purposes, marine operations, and for general planning and management of the economy. There is need to increase the accuracy of such forecasts through better training of personnel and the acquisition of state-of-the-art facilities.

THE NEED FOR TRAINING IN SPACE SCIENCE AND TECHNOLOGY

To sustain an efficient and broad based application of space technology in Kenya will require a critical mass of trained personnel in various categories of the technology. A proper take off and continued growth of space industry in the country, will require adequate training and specialization in the applied and pure science aspects of the technology.

The Kenya space science policy, which was prepared under the auspices of the National Council for Science and Technology (Aligula et al., 1994), recommended the setting up of a Kenya Space Agency (KSA) to be charged with the responsibility of designing and coordinating the appropriate training programmes in space science. This, in turn, would facilitate the implementation of techniques and procedures in order to: aid economic planning and enhance international co-operation, maintain and improve domestic and international communication services, facilitate natural disaster prediction and mitigation efforts, improve the dissemination of meteorological information, establish a national survey control network and aid the establishment of deformation monitoring schemes to ensure the safety and continuous operation of engineering structures.

A proper implementation of the designed space science training programmes would play a major role in re-shaping and re-orienting the country's scientific goals. This consequently would lead to a better understanding and manipulation of space science potentials towards socio-economic development, environmental protection, promotion of regional and international co-operation and international peace and security.

During the United Nations (UN) regional conference on space technology for sustainable development in Africa, held in Dakar, Senegal, 25th-29th October, 1993, it was noted that there were insufficient space research and training institutions, satellite launching and tracking stations, communication satellites and remote sensing ground receiving stations (there is only one situated in Johannesburg, South Africa). On realizing this regrettable situation, the UN urged African Governments to locally make adequate long term commitments and investments towards the development of skills and knowledge as well as the required infrastructure in Earth observations and related disciplines. In order to implement this UN call, Kenya will require rigorous and well planned training in space science at the technician level and for scientists and engineers. This can be achieved through concerted Government efforts coupled with financial and technical support from the international community. Such government efforts need to include the initiation of and full participation in joint ventures in space science projects.

The Kenyan space policy document of 1994 points out that the limited application of space technology in the country has been, to a large extent, due to the lack of sufficient base of qualified personnel. Although there is a reasonable number of Kenyans who have been trained in certain aspects of applications of space technology, such as telecommunications, meteorology, remote sensing, positioning, environmental monitoring, mapping and deformation monitoring, there are no experts who have specialized in space science per se. Furthermore, there are no facilities for undertaking space science studies locally. In fact there are no dedicated and coordinated training programmes in space sciences in Kenya. The public universities have, in their programmes, some aspects of applied space sciences which are limited in scope and are adversely affected by the lack of adequate facilities.

Kenyans with knowledge in the applied aspects of space sciences have been trained abroad. Generally, such training has not been purely on space sciences but to a large extent in other disciplines in which knowledge in a particular space science may be needed for completeness and adversity. This situation, together with the lack of reasonable level of trained manpower in the space science, has greatly contributed to the slow progress of space technology in Kenya.

Therefore, there is a clear need for coordinated programmes for space science training. This will require that the existing facilities for training be expanded and upgraded to facilitate specialized training in telecommunications, remote sensing, surveying, hydrography, deformation and environmental monitoring and meteorology. In this regard, it is also critical that international co-operation be enhanced and joint ventures in space science projects be initiated to supplement training requirements and provide exposure to modern facilities which may be available only in developed countries.

In 1990, a sub-committee of the Kenya steering committee on the San Marco project was formed to deliberate on training on space science research and technology and draw-up a training programme in space science (Ogana et al., 1990). In their report, four categories of personnel were identified namely:- Technicians, Technologists, Aerospace Engineers and Scientists and basic and applied scientists. It was noted that only a few technicians have been trained under the San Marco project. The subcommittee further observed that Kenya can only take a significant step towards space technology and reap maximum benefits from it by giving priority to training of Technologists and Aerospace Engineers and scientists who, subsequently, would be able to: (i) fully participate in the launching, tracking station, (iii) decode and synthesize data from satellites and disseminate such data to potential users, such as the basic and applied scientists, and (iv) design and manufacture satellites and spacecrafts and their payloads.

Although Kenya seriously needs to improve her socio-economic development through the application of space technology, she lacks the financial capacity to undertake these advanced space science training initiatives on her own. In order to realize these goals, she requires co-operation from the international space science community in terms of financial support, technical support and joint ventures in space science projects.

CONCLUSIONS AND COMMENTS

The paper discusses the background of the application of space technology and the impact which the technology has had in Kenya. The role that space science can play and the opportunities available in Kenya are spelt out. It is noted that, despite the high potential of the technology, its impact in the country remains small, largely due to the lack of a critical mass of trained personnel to spearhead a proper utilization of the technology.

The lack of specialized training in space science and technology is a major handicap to the growth of space industry in Kenya. For space technology to make a viable economic and social impact, it is critical that expert training programmes be undertaken as a prerequisite. In this regard, it is to be noted that Nairobi houses several international organizations which may be called upon to assist in the drawing up and coordinating appropriate training programmes in space science. These organizations comprise of the United Nations Agencies like: UNESCO, UNEP, HABITAT, the Regional Centre for Services in Surveying, Mapping and Remote Sensing (RCSSMRS), etc.

It is recommended that concerted efforts be made by the Government of Kenya and the international space community to invest in space technology in the country and to formulate and implement appropriate space science training programmes. This, consequently, would enhance and sustain the growth of the space industry, guarantee international peace and security as well as boost and modernize parameters for socio-economic development.

International co-operation in space technology is responsible for the limited achievement which Kenya has made in the application of the technology. Such cooperations ought to be expanded, particularly in areas like surveying, mapping and hydrography, where the application of the technology has been minimal.

A recent example of such co-operation was the launching of a "hightech link" between Kenya Broadcasting Corporation (KBC) and the Deutsche Welle Television of Germany on March 28, 1996 in Nairobi. The hightech link allows KBC to decode and broadcast Deutsche Welle TV programmes. During the launching, KBC obtained a satellite receiver which will enable the corporation to select (for their viewers) from a wide range of programmes broadcast on Deustche Welle TV. The co-operation between KBC and Deutsche Welle TV on telecommunications has been in existence for the last three decades, having started in 1963. During the period, more than 30 employees of KBC have been trained at Deutsche Welle station in Cologne. Similar co-operations are urgently needed in order to evolve the necessary team work required to expand the impact of space technology in Kenya.

Furthermore, the launching of the East African Co-operation by the Presidents of Kenya, Uganda and Tanzania, on March 14, 1996 has added some impetus and created a sense of urgency in the application of space science and technology in the region in order to assist in the integration and rapid growth of the region's economy and social structures. The East African Co-operation has created a larger market of 70 million people in the region and other bordering countries like Burundi have expressed their intentions to join the co-operation in the near future. This is a challenge to the international space science community to invest in the region and support the training of experts to ensure a rapid growth of the application of space technology in the region.

REFERENCES

Aligula H.M., Ogallo L.A., Aduol F.W.O., Aseno J.O., Obel J.D. et al. (1994). "Kenya space science policy" National Council for Science and Technology, June 1994, Nairobi, Kenya.

Ogana W., Kola B.O., Aduol F.W.O. et al., (1990) "Training in space science research and technology". A report by a subcommittee of the steering committee of the San Marco Project, November 1990, Nairobi, Kenya.

Pergamon

Adv. Space Res. Vol. 20, No. 7, pp. 1411–1419, 1997

Printed in Great Britain
0273-1177/97 $17.00 + 0.00

PII: S0273–1177(97)00740–0

SPACE SCIENCE EDUCATION IN THE AFRICAN CONTINENT

J. O. Aseno

Department of Surveying, University of Nairobi, P.O. Box 30197, Nairobi, Kenya

ABSTRACT

Through measurement and interpretation of the spectral, spatial and temporal variations in electromagnetic emissions and reflections from the Earth's surface, important information related to natural resources can be acquired. Furthermore, satellite technology has greatly improved the communication and positioning techniques world-wide. Consequently, space science now provides valuable and timely information about natural resources, which has become a major factor in sustainable development.

The realization of the full potential of space science in the context of development in Africa requires adequate education and training in order to facilitate project formulation, planning, management and implementation. This, in turn, would lead to the formulation and adoption of national space science policies based on user needs and addressing both the short and long-term needs of a particular country.

Space science education in Africa needs to address issues like (i) provision of programmes, (ii) integration of the proposed techniques within the existing infrastructure, and (iii) training in Remote Sensing, Global Positioning System, Geographic Information System and other space science techniques, in order to ensure the successful implementation of space science projects within the continent. In this context, African universities ought to play a major role in space science training, research, consultancy and publication. Through international co-operation, it is possible to develop and support national, regional and international training programmes and international scientific exchange in Africa.

INTRODUCTION

Recommendation number fifteen of the United Nations (UN) Committee on the peaceful uses of outer space (1994), in the report on the United Nations regional conference on space technology for sustainable development in Africa, held in Dakar, Senegal, 25-29 October, 1993, prepared by the Scientific and technical Sub-Committee, 31st session, Vienna, Feb 21 - March 4, 1994, states that: "Different aspects of space science and technology, including its fundamentals, development and related applications should become integral part of appropriate curricula at African Universities and technical Institutions. These should be initiated and enhanced with a view to encouraging African scientists to actively participate in and contribute to the understanding, evolution, growth and application of this technology in the national and regional social and economic development programmes. On request, the United Nations, through its space applications programme, can assist in developing the necessary curricula in remote sensing, satellite meteorology, satellite communications and basic space science as may be appropriate in each case. This recommendation should be transmitted to all African Universities and research centres through the Association of African Universities".

The above UN recommendation is strongly supported and ought to be fully implemented since local Universities in the African countries have a key role to play in the increased application of space technology in the continent through education, research, training, consultancy and publication. What is urgently required is the increase of capacity of such Universities in order for them to fulfil such objectives through (i) introduction and/or expansion of undergraduate and postgraduate education and training curricula in space science and technology in selected established Universities in Africa, (ii) training in the various principles and applications of space technology, (iii) support and financing of research in space science and technology at University level, and (iv) publication of research findings.

Space technology has the potential to provide valuable information about natural resources, mapping and positioning, which are essential in sustainable development. Despite its potential, the technology has not been adequately implemented in Africa. Most African countries are constrained by the lack of communications, up todate resource and land maps, computer knowledge and training.

The advancement of space science in Africa will require specialized knowledge and, increasingly, advanced computer technology. Consequently, the improvements in cost-effectiveness, accuracy and precision of information products from space science projects in Africa will be intimately tied to adequate training of African personnel, who in turn, would be able to participate fully in the design, planning, execution and management of such projects.

The paper points out that there is an urgent need for scientists and engineers from Africa to be involved in the planning of future space science missions in order to strengthen local capacity building as well as to ensure that they meet the expanding space science needs in the African countries. To achieve this will require prior education and training. Funding should be made available from donor and space science agencies to facilitate this involvement. It is further suggested that space science education in Africa should be introduced in all relevant undergraduate and teacher training curricula and postgraduate programmes in space science should be offered in the larger African Universities, while other specialized training should be undertaken abroad.

Priority areas in training should include the principles of various space science techniques involving (i) short courses, and (ii) distance learning packages for public, private, non-governmental organisations as well as military sector personnel. Since education and training leads to technology adoption and use, it would be essential to achieve this in the three categories of basic training, on-the-job training and hardware and software maintenance. It is finally argued that appropriate education and training in space science in Africa would make it possible for governments and decision-makers to benefit from space technology through international co-operation, the identification of the country's priorities and needs, the formulation of national policies and by subsequent sustainable development.

THE STATE OF SPACE SCIENCE EDUCATION

Surveying

The recent advances in space technology (e.g. Global Positioning System, GPS) have greatly revolutionised land surveying worldwide. Unfortunately, GPS, which is a major space technology for Surveying, has had little impact in Africa despite the fact that this technique has been operational since the mid- 1980s. The lack of impact has been due largely to inadequate education and training of personnel and the cost of applying the technology, which appears to be out of reach of most African countries. The problem is further compounded by the fact that there are very few Universities in Africa which offer degree programmes in Surveying and the component of space technology in such programmes is also inadequate. In the Anglo-African region, such Universities are only found in Nigeria, Kenya, Zimbabwe and South Africa.

Furthermore, there are very few space science education centres dealing with land Surveying. The only such centres are the Regional Centre for Services in Surveying, Mapping and Remote Sensing (RCSSMRS) in Kenya, Photogrammetry and Remote Sensing Centre in Nigeria and the Wagadugu Centre in Burkina Faso. Therefore, there is need to increase and expand such centres as well as to expand curricula of Departments of Surveying in appropriate Universities in order to ensure adequate training for the full application of space technology in Land Surveying and Marine geodesy.

A continental GPS project is urgently required in Africa for the establishment of the African datum, providing a uniform continental network which would be used as (i) a framework for conducting studies in crustal and other Earth movements in order to monitor the occurrence of natural disasters in Africa, and (ii) a datum for lower order Surveys (cadastral, hydrographic, engineering, etc.) and resource mapping, which would yield (a) underwater maps to facilitate marine navigation and delimitation of cadastral boundaries, and (b) maps of agricultural lands, forests, game parks, arid and semi-arid areas, etc, for economic planning.

Education and training in Land Surveying related space technology may be achieved through the combined effort of the manufacturers of the relevant space technology equipments, Government agencies and private organisations. This would lead to a proper identification of appropriate and well targeted projects as well as combined marketing and commercialization of the products.

Such adequate education and training is urgently required in Africa in order to promote the various areas of application of space technology. This, consequently, would enhance socio-economic development. Application areas where tangible results would be achieve rapidly include health education and environmental awareness by satellite technology through television. Equally to be addressed at this point will be the need to provide incentives after training in order to reduce the current brain drain from Africa.

Remote Sensing

The level of application of remote sensing in Africa is generally low due to insufficiently trained remote sensing operators. This has created a huge gap between the demand for and supply of remote sensing products.

Most African Universities have not appreciably incorporated remote sensing curricula in their academic and research programmes. Only service courses, which are limited in scope and duration, are undertaken, largely by technical Government institutions and international institutions like the RCSSMRS. However, this trend is beginning to change. For example, the University of Zimbabwe is beginning to make an impact in the education and training in remote sensing and, in some South African Universities, Masters of Science programmes in remote sensing are now being offered.

Remote sensing centres which provide short courses to the lower cadre personnel are the RCSSMRS in Nairobi, Kenya, Regional Centre in Wagadugu, Burkina Faso, Ile-Ife University in Ibadan, Nigeria, and national institutions within the appropriate Ministries in Kenya, Egypt, Morocco, Zaire, Tunisia, etc. Recently, additional two centres have been approved for establishment in Nigeria for Anglophone Africa and Morocco for Francophone Africa.

It is to be emphasized here that instructors who could help in developing remote sensing technology in Africa in terms of the design of sensors, processing of techniques, and on-the-job training of low cadre personnel are seriously in short supply. Training of such instructors require at least a two year postgraduate programme in a University with appropriate facilities. The training of such instructors is critical for the future expansion of remote sensing activities in the continent. This would lead to the launching of high resolution remote sensing satellites [like the South African green satellite (ground resolution 5m), and the Indian satellite (ground resolution 5m)] in several regions in Africa.

To facilitate the proper application and expansion of remote sensing technology in Africa, adequate University level education and training is required. This could be achieved through expansion and improvement of facilities in regional Universities to cater for the required training in the continent. Possible Universities and the regions they would cater for are as follows: Moi University, Eastern Africa; University of Zimbabwe, Central Africa; University of Cape Town, Southern Africa; University of Morocco, Northern Africa, University of Ibadan, Western Africa and University of Gabon, Central Africa.

The number of African remote sensing experts with Masters of Science qualifications are very few. For example, there are only about ten such people in Kenya. This number is probably lower in any East and Central African country since, in this region, Kenya was the first to institutionalize remote sensing. In the Francophone African countries (e.g. Algeria, Tunisia, Morocco, etc.) the number is slightly higher due to a higher involvement and support from the French Government.

Adequate education and training in Africa can easily be achieved through: (i) sensitizing the decision makers and making them fully aware of the benefits of enhanced training programmes. Such decision makers must be prevailed upon to attend and participate in appropriate seminars and workshops in order to fully understand and appreciate the potential of remote sensing, (ii) initiation of national ministerial and private training and research centres to train low cadre personnel, provide on-the-job training, generate ideas and create jobs and (iii) introduction of remote sensing in high schools in order to instil background knowledge and understanding in the rural areas.

Generation of funds to sustain remote sensing activities may be possible if appropriate Government Ministries were to provide "seed-money" to allow private companies and Government agencies to market and commercialise remote sensing products at affordable fees to the targeted customers. Examples of such products include (i) value-added images to assist in solving Earth resource problems, (ii) thematic maps e.g. tourist maps, township maps, land-use maps, (iii) teaching materials, and (iv) land registration maps.

Therefore, increased education and training in remote sensing in Africa will result in improved socio-economic development through better knowledge, exploitation and protection of resources; expanded initiation and implementation of joint ventures in remote sensing projects; and exchange of experts to facilitate assessment, evaluation and improvement of the performance of remote sensing projects.

Telecommunications

In Africa, there are few people who can fully comprehend space telecommunications and exploit its full potentials. Institutions like the post offices normally operate in specialized and limited areas and are therefore hindered by the lack of know-how in the recent advances in space technology. General developments and improvements in space telecommunications will require on-the-job training, capital investments, co-operation between Governments and foreign embassies, non-governmental organizations, etc. Due to the lack of a critical mass of trained personnel in most Africa countries, national and international telecommunication networks are weak. This, subsequently hinders national, international and regional exchange of information. This adversely affects businesses associated with telecommunications, thereby retarding national and continental economic growth.

A paper presented by the Kenyan Minister for Information and Broadcasting to the Information Society and Development Conference (ISAD) held in South Africa, May 13-17, 1996, revealed that Africa has less than 2 percent of the world's telephones lines. The Minister further pointed out that while the global telecommunication sector was changing drastically in response to a changing global economy which is relying on advanced information and telecommunications technology, the situation in the developing countries remains deplorable.

However, it is to be noted that, despite the general inadequate telecommunication networking in the continent, some Africa countries (e.g South Africa, Egypt, Libya, Ivory Coast, Morocco, Nigeria, Algeria, Kenya, Congo, etc.) have done relatively well in the development and implementation of space telecommunication techniques. Good space telecommunication training programmes are found in some African Universities, e.g. in South Africa and Nigeria, and nearly all African-Arab countries send their people to European Universities and other Universities in developed countries, to acquire relevant education and training in space science and technology.

Meteorology

The level of education and training in meteorology is generally low in Africa. However, the Francophone African countries are ahead in terms in trained personnel. Other countries that have done relatively well in terms of education and training include Kenya, South Africa and Egypt. The meteorological training school in Dagoretti, Nairobi, Kenya, caters for Eastern and Central African countries. Most of the financial support for meteorological activities in Africa comes from the World Meteorological Organizations (WMO) whose regional office for Africa is in Nairobi, Kenya. Despite this WMO financial support, there is still a serious lack of specialization and diversification in meteorological education and training in Africa.

There is a regional telecommunication hub in Nairobi through which all meteorological data transmitted to and from Africa pass. In addition, the MWO regional office in Nairobi has recently acquired the latest telecommunication equipment from the French Meteorological Services, which speeds up the process of data reception and dissemination in Africa.

NEED FOR AN EXPANDED SPACE SCIENCE EDUCATION AND TRAINING

Education and training leads to technology adoption and use [Catizzone and Fanta, 1993]. An increased use of space technology in Africa requires specialised knowledge in the various disciplines of the technology, backed with advanced computer technology. Higher education in these two areas is critical for the advancement and application of the technology in the African continent.

Advances in space science education in Africa will lead to improvement in the design and planning of projects, thereby resulting in higher accuracies and precision of information products in space science activities. This, in turn, will facilitate the realization of socio-economic benefits of space technology. In addition, the establishment of multidisciplinary space science committees, composed of adequately trained citizens, will greatly assist in preparing guidelines and policies for identifying priority areas to aid individual countries and the donor agencies in designing, implementing and managing new projects.

Since education and training are essential components of the sustainable application of space technology, a rapid take-off of the space industry in Africa will require that: (i) initiated space technology projects should also include training opportunities for the African personnel, (ii) more facilities for education and training be put in place in different countries in Africa, (iii) an inventory of existing educational and training opportunities in Africa be made in order to assist in assessing the level of expansion required in different areas, (iv) space agencies promote on-the-job training whenever new satellites and space technologies are introduced, and (v) sponsorships for further studies in space science be made available for all levels of training in foreign countries in which better training facilities may be available.

According to Billingsley and Strome (1986), the use of digital imagery and other remotely sensed digital data has grown rapidly in many disciplines in the last three decades. This rapid growth has been possible because of parallel developments in digital image education and acquisition, advanced computer processing capabilities, enlarged storage and retrieval capacities, transmission and dissemination link capacities and improved image display technology. The increasingly widespread availability of data has encouraged the investigation of world size needs for training and education in order to improve the application of the globally available space science data in a third world continent like Africa.

Consequently, there is an urgent need for adequate education and training of the African scientists and engineers in order that such scientists and engineers may subsequently take a leading role in planning and designing future space technology activities and projects to ensure that such activities and projects fulfil the socio-economic priorities of their respective countries. For example, in order to achieve effective planning in remote sensing, the African countries need to express their requirements in the framework of the Global Ocean Observing System (GOOS), the Global Terrestrial Observation System (GTOS) and the Committee on Earth Observation Satellites (CEOS). Financial and technical support from the donor agencies ought to be expanded to facilitate this involvement.

It is observed from the Commission of the European Communities (1996) document on the analysis of constraints and opportunities for cost-effective implementation of Earth observation techniques in developing countries that space technologies have developed faster than institutions and attitudes. A radical approach is required in order to take advantage of the high potential of space technology. In Africa, assistance is required in the process of technology transfer as well as in structuring their information systems and decision making processes in order to achieve full benefits of space technology. This can be realised through improved and well focused education and training of the African personnel.

Awareness of environmental issues and use of space technology differ considerably in the African countries due to the lack of adequate know-how, poverty, drought, etc. Therefore, increased education in space science would greatly assist in improving socio-economic growth, thereby reducing the environmental awareness gap through the use of digital mapping system (including interactive computer graphics systems and geographic information systems) for the creation of digital cartographic, topographic and land related data bases. With the recent advances in space technologies, digital mapping systems have become essential in education and training programmes, environmental monitoring and protection as well as resource surveys.

The Commission of the European Communities (1996) further reveals that in developing countries, "many national information networks comprise a variety of observation, measurement and monitoring activities of variable quality, quantity, efficiency and utility. They are rarely linked to each other or coordinated more than nominally. There is often a considerable overlap of mandate and activities, yet little of the derived information is circulated and even less is well used. Decision makers are frequently ill-served with little up-to-date information and have learned to get by without it. The consequence is that most decision makers are in a state of uncertainty most of the time, and less effective decisions are made. This presents a magnificent opportunity for major benefit to be obtained from appropriate utilization of cost-effective information technology and timely information sources".

In terms of the application of space technology in Africa, this unfortunate scenario may be turned around through increased and well focused space science education and training. Such education and training have the best chance to improve the level of information organization (i.e. acquisition, dissemination, storage retrieval and accessibility) of space science data for effective decision making, resource and environmental management and other related socio-economic developments.

PROSPECTS AND BENEFITS OF ENHANCED APPLICATION OF SPACE TECHNOLOGY

An expanded and well focused space science education in Africa will subsequently lead to the establishment of competent national agencies in space science to spearhead national interests in space science as well as to initiate and coordinate international ventures in space science training, project formulation and implementation. This would culminate in the development of well considered space science policies within the continent. Such policies would be determined by the national agencies in conjunction with the representatives of the funding organizations, research councils, learned societies, non-governmental organizations and industry.

These national agencies would be charged with the task of co-operating amongst themselves in Africa in order to:

(i) provide education and training facilities in space science in order to ensure continued adequate education and training of the African personnel in the various fields of space technology;

(ii) promote the acquisition, supply and exchange of space science data in order to facilitate a coordinated and well managed application of the technology.

iii) assist the space science industries, in individual countries, with the planning, design, implementation and management of national and private space science projects;

(iv) provide facilities for research in and development of image processing techniques, satellite positioning, crustal deformation monitoring, using satellite techniques, telecommunication linkages, hydrographic mapping, etc;

(v) act as national focal points for the coordination of education and training and the development of space science techniques and their applications.

Within the above framework, it is anticipated that the potential benefits of space technology can adequately be exploited in Africa. This is possible because there is a lot of goodwill within the international community to co-operate in space science activities. For example, the earlier cited Commission of European Communities (1994) document recommends that the commission works together with member states, United Nations agencies, international institutions and other donors, to assist developing countries structure and evolve cost-effective basic national information networks towards meeting operational needs for information by all concerned. This indicates a clear willingness by the Commission to co-operate in space science ventures and therefore becomes a major prospect towards the enhancement of the application of space technology in Africa. In this regard, the European Community and other relevant organizations, need to be approached to assist African countries in the planning and implementation of appropriate education and training programmes that would consequently provide a strong basis for the increased application of space technology in Africa.

Through international co-operation, it would be possible, for example, to use a combination of radar and satellite imagery in real time to provide better forecasting in terms of the estimation of the probability of rainfall. Eccleston et al. (1982) observe that radar and satellite data can easily be overlaid on a colour television display to show the areas of cloud which are associated with surface precipitation. Since television sets are increasingly finding their way into the African countries, initiating and implementing a weather forecasting training programme using television sets would achieve tangible results towards increased application of space technology in Africa.

Furthermore, the usefulness of air-and space-borne Synthetic Aperture Radar (SAR) imagery for geoscientific applications can be improved by the integration of various radargrammetric processing methods. Buchroithner and Triebnig (1986) have found out that radar images display ground range non-linearity and along-tract skewness induced by radar mapping characteristics and residual SAR system biases. This, in turn, affects both the geometric and radiometric quality of the data. Therefore, the generation of synthesized SAR imagery by simulating radar projection and target response would allow for systematic investigations into the problems of land extraction and radar stereoscopy. The application of such space techniques would greatly enhance resource mapping and environmental monitoring and management in Africa.

In summary, an enhanced application of space technology in Africa will:

(i) facilitate the revision and improvement of Government strategies and policies towards socio-economic development;

(ii) improve efficiency and effectiveness in the activities of the non-governmental organizations;

iii) lead to the expansion and better coordination of the national and regional space science education and training facilities;

(iv) lead to the expansion and improved effectiveness of United Nations activities in Africa;

(v) provide a chance to realise sustainable socio-economic development, thereby reducing the dependency on the international donors (e.g. International Monetary Fund, World Bank, etc.) to finance economic and social programmes in Africa;

(vi) lead to refinements and improvements in environmental monitoring and protection techniques so as to effectively reduce pollution and the degradation of the environment in general;

vii) reduce poverty and war in Africa through provision of better farming methods and pest control mechanisms; improved industrial growth resulting from better resource surveys and resource exploitation techniques; and increased exchange of information, trade and other joint economic ventures, thereby resulting in regional and international co-operation and social co-existence.

CONCLUSIONS AND COMMENTS

Despite the vast amount of space data currently generated globally through the use of artificial satellites, most African countries still have poor space science information networks for resource and environmental management. Due to the lack of adequate education and training in space sciences, the capacity of the African scientists and institutions to process space science data and to disseminate those data remains dismally low. The result of this is that the socio-economic planning processes in many African countries tend to be out-of-date, under-resourced and unevenly implemented in areas which such implementations are attempted. The consequence of this is that the decision making process is not adequately improved and the economic growth is hampered.

This situation clearly illustrates the need for a better focused and well structured space science education and training in Africa in order to improve decision making towards sustainable resource management and socio-economic development. It is a considered opinion that the solution lies in the establishment and expansion of regional and national space science education and training centres with clear objectives and well-defined management frameworks.

It is to be observed that training priorities and general issues differ greatly from one country to another in Africa. Therefore, within the framework of the proposed education and training centres, it would be necessary to look at these issues deeply and point out areas that require particular levels of training. This, in turn, would assist in identifying the priorities of individual countries in terms of the application of space technology.

In view of the underdevelopment of multi-level infra-structures and networks in space science in many African countries, it is considered vital that (i) networking capacities be improved in order to link experts in nearby countries so as to facilitate the exchange of information, and (ii) the functions and coordination of regional centres be strengthened through advanced education and training of personnel and the establishment of additional infrastructure (e.g. data acquisition facilities) to ensure efficient communication and transfer of information.

The proposed networking strategies require, as a pre-requisite, international cooperation and policy towards space technology in order to encourage the participation of the African countries in the identification and design of regional and international space science education activities. This, in turn, would lead to the enhancement and expansion of the educational role of the national, regional and international centres through technical and financial support of space science training programmes.

One of the key resolutions of the UN Africa GIS '95, held in Abidjan, 6th to 10th March, 1995, under the theme "Geographical Information Systems in support of sustainable management of natural resources in Africa" was (i) that an effective long term training programme should focus on building capacities within national and regional training centres and on coordinating training activities between regional centres, Universities, governmental organizations and other users, (ii) that too often in Africa, the importance of the national university as a stable source of on-going training has been overlooked, and (iii) that donor support should focus on building complementary capacities within these organizations.

It is vital that such resolutions be consolidated with the issues and recommendations discussed above in order to facilitate the development and implementation of well designed training and education programmes in space science and technology. This would consequently make it possible to achieve tangible results in the application of space technology in Africa within a reasonable period of time.

REFERENCES

Billingsley F. and Strome W., (1986). "Standardization of remote sensing and GIS data transfer". Proceedings of the symposium on photogrammetry and remote sensing systems for data processing and analysis, May 1986, Maryland, U.S.A. American Society of Photogrammetry and Remote Sensing, pp 182-189.

Buchroithner M. and Triebnig G., (1986). "Radargrammetry as a tool for geoscientific applications". Proceedings of the symposium on photogrammetry and remote sensing systems for data processing and analysis, May 1986, Maryland, U.S.A. American Society of Photogrammetry and Remote Sensing, pp 322-332.

Catizzone M. and Fanta A., (1993). "Needs and prospects of remote sensing in developing countries". A workshop supported by the European Commission-Directorate General XII for science, research and technology, the European Space Agency and the Regional Centre for Services in Surveying, Mapping and Remote Sensing, held in Nairobi, Kenya.

Commission of the European Communities, (1996). "A summary of the analysis of the constraints and opportunities for cost effective implementation of Earth observation techniques in developing countries". Space for our World, ETES-93-0042, January 1996.

Eccleston A.J., Carpenter K.M. and Collier C.G., (1982). "The use of meteosat imagery and a network of radars for observing and forecasting rainfall". Proceedings of the Annual technical sensing and the atmosphere, held in Liverpool, December 1982. Remote Sensing Society, pp 22-28.

United Nations Committee on the peaceful uses of outer spacer, (1994). "Report on the United Nations Regional Conference on Space Technology for sustainable development in Africa, held in Dakar, Senegal, 25-29, October, 1993". Scientific and technical Sub-committee, 31st session, Vienna, February 21-March 4, 1994.

Pergamon

Adv. Space Res. Vol. 20, No. 7, pp. 1421–1425, 1997

Printed in Great Britain
0273-1177/97 $17.00 + 0.00

PII: S0273–1177(97)00741–2

SPACE SCIENCE EDUCATION AND TRAINING IN BRAZIL

M. A. Abdu and I. S. Batista

Instituto Nacional de Pesquisas Espaciais-INPE, C.P. 515, 12201-970-São José Campos, SP, Brazil

ABSTRACT

With the beginning of space research in Brazil in the early 60's a national commission for space activities was created. This commission took important steps, in parallel, to create a program of education and training activities that was forseen as of critical importance for the maintainance of a desirable to ambitious degree of space activity in the country. Early efforts were, as a first step, in favour of training of Brazilian students in foreign centers of excellence (mostly in the US and Europe), as a means for achieving the potential human resource necessary for eventual development of a national program. Today a large percentage of the needs in the area is met by programs developed inside the country at different institutions, the most notable being the "Instituto Nacional de Pesquisas Espaciais" -INPE (National Institute for Space Research) at São José dos Campos, where post graduate courses and training programs are well established. Under-graduate courses in selected areas are provided at some Universities, notably at the University of São Paulo. A brief description of the existing programs and an evaluation of the results being achieved are presented in this paper.

INTRODUCTION

Space science education and training constitute the basic foundations for the growth of organized research and development activities in this field. In Brazil the space research activities had its start in the early 60's with the creation of a commission for the coordination of a national space activity (GOCNAE). The early focus of research was on the earth's ionosphere, magnetism and upper atmosphere, involving use of ground based diagnostic instruments such as ionosonde, magnetometer and laser radar. With the transformation of GOCNAE into INPE in 1971 diverse activities were started in space application and technology areas. Today the activities in space related areas are distributed at different locations, under INPE's administration, as well as in some universities, in Brazil owing to the educational and training program initiated in the years following the start of the space research program. An overwhelmingly major part of the educational and training activities in the field in Brazil is being organized by INPE, and therefore the following brief overview of the these activities are based mostly on the experiences of INPE for over the last three decades. Initial efforts were, as a first step, in favour of training of Brazilian students in foreign centers of excellence (mostly in the United States and Europe) as a means for achieving the potential human resources for eventual development of a national program. The first group of students was recruited in 1967, majority of them being in electrical engineering and physics areas. These researchers completed their Masters degrees at INPE and Ph.D degrees abroad, and on their return to Brazil constituted part of the initial nucleus for research and training activities of INPE. The nucleus also consisted of, from the beginning, experienced researchers recruited by INPE from abroad (notably from India, Japan, UK and USA). The post-graduate program started producing MS and Ph.D degree holders starting from 1968 and 1980 respectively, and these graduates (mostly of Ph.D degree) now belong to the nucleus of the space science education and training activities of INPE and other centers (including Universities) in Brazil. In the following we will present brief descriptions of the INPE's post-graduate and short term training programs and a mention about the under-graduate and post-graduate courses in some areas available in Brazilian universities.

POST-GRADUATE PROGRAM AT INPE

The post-graduate program at INPE offers courses in the following areas: space geophysics, astrophysics, meteorology, remote sensing and space technology and engeneering (with emphasis on space mechanics control, and in combustion/propulsion). The brochures periodically published by the institute (in Portuguese language) provide detailed information, including the curriculum of these different courses. Table 1 provides the number of teaching and research staff available for the different courses.

Table 1. Teaching Staff for the Different Courses

COURSE	PROGRAM	NUMBER OF TEACHERS HOLDING	
		PhD	MS
Space Geophysics	MS & PhD	34	-
Astrophysics	MS & PhD	13	-
Meteorology	MS & PhD	21	3
Remote Sensing	MS	38	2
Space Thechnology and Engineering/ Combustion and Propulsion	MS	6	3
Space Technology and Engineering/ Space Mechanics and Control	MS & PhD	15	-

Fulfilment of number of credits on basic and optional subjetcs, qualifying exam and completion of a thesis on a relevant research topic are the minimum requirements for the award of the MS and PhD degrees. Since its start in the 60's the post-graduate program has undergone periodic reorganization in recognition of such factors as the evolution of the teaching staff, creation of new research and development projects and the academic requirement for each course demanded by the Brazilian Federal Concil for higher educational personal (CAPES), that evaluates the post-graduation degrees in Brazil. The number of students opting for the different courses is decided by the job potentials and demands in the industries, universities and other sectors. Thus there is a large percentage of the incoming students admitted in application oriented areas such as remote sensing and meteorology as compared to space geophysics and astrophysics areas. Students are admitted to INPE's post-graduate program from universities all over Brazil and from neighbouring countries, especially, Argentina, Bolivia, Chile and Peru.

A statistical representation of the students admitted to masters degree between 1988 and 1995 is presented in Figure 1. The top panels show the number of students admitted for different courses, the middle panels the number of students who completed the degree and the bottom panels the statistics of the discontinued students. It may be observed that the duration for completing the Masters degree varies from 2 years for meteorology to ~3 years for remote sensing. The percentage of students who discontinue seems to differ from one course to the other, the smallest being that of remote sensing. Completion of the number of Masters degrees in all fields produced by INPE's post-graduate program is presented in Figure 2. There is a clear growing trend in the number of qualifying MS graduates with large variabilities from one year to the other.

Similar statistics as that of Figure 1 are presented for PhD degree in Figure 3. It may be noted that Ph.D program is not offered for remote sensing and combustion/propulsion areas and that the number of years required to complete PhD degree varies from 5 years in space geophysics and astrophysics to 6 years in meteorology. A composite of statistics for the MS and PhD students completed in three broad areas (grouping together space technology and engineering space geophysics and astrophysic) starting from the first MS degree in 1968 is presented in Figure 4. The first PhD's were completed in 1981 in space geophysics and meteorology areas. It may be observed that the numbers of both the MS and PhD graduates have maintained an increasing trend since the inception of the programme.

Students completing MS degree from INPE could opt to pursue further studies in universities abroad, especially, in areas where facilities are lacking in Brazil. INPE also encourages its PhD graduates to acquire post doctoral training in the field of their PhD specialization in leading centers abroad. The statistics of INPE students who complete PhD and post doctoral research abroad are presented in Figure 5.

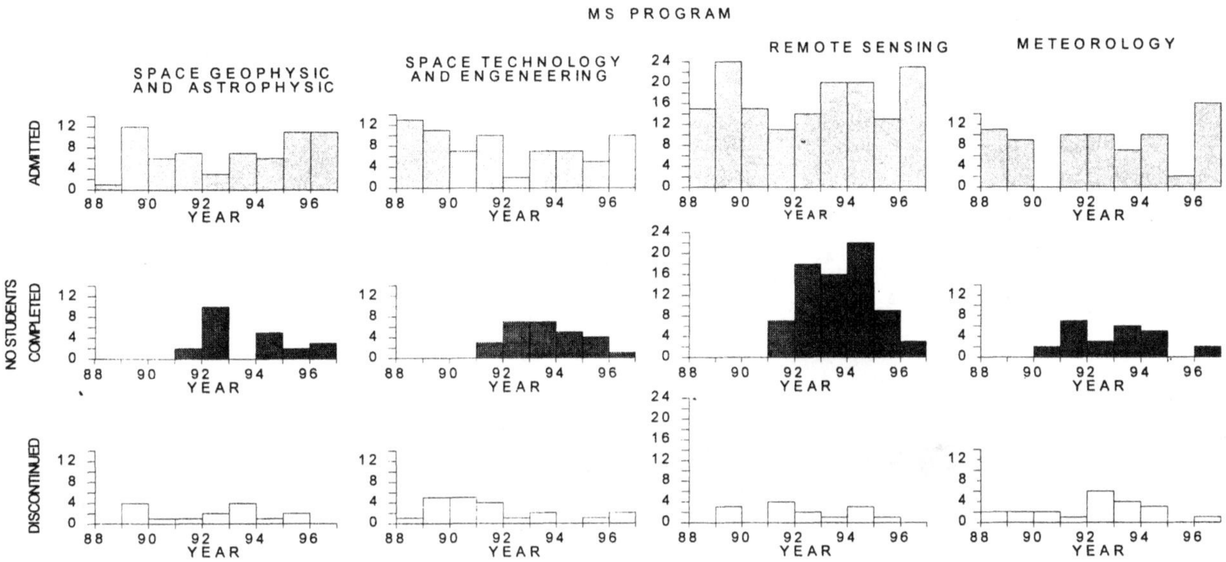

Fig. 1. Statistical representation of students admitted to Masters degree at INPE.

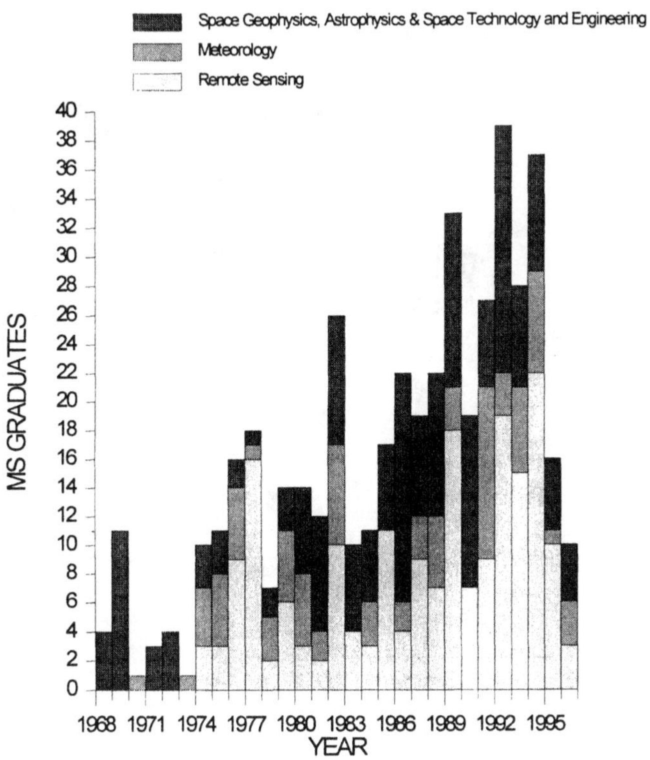

Fig. 2. Number of Masters degrees produced by INPE's post-graduate programs.

SHORT TERM TRAINING COURSES

INPE organizes training programs to qualify specialists in remote sensing area from developing countries, especially, from Latin America and Africa. A series of 9 such courses organized since 1985 at São José dos Campos was in collaboration with United Nations Outer Space Affairs Division, the United Nations University, the European Space Agency, the coordination for the improvement of Higher Education Personnel (CAPES) and the National Council for Scientific Research and Development (CNPq). The course objectives were:

- to qualify specialists from developing countries in remote sensing research and techniques for survey and monitoring of natural resources;
- to increase the use of remote sensing techniques related to survey, planning and management of natural resources in developing countries;
- to identify the remote sensing utilization in the field of specialization and work situation at home;
- to organize and collect information from remote sensing data to be used in their planning and decision making procedures; and
- to recognize the increasing contribution of remote sensing to analyse environmental impact that may impair the development of these countries and indicate measures or further investigations to avoid the side effects of these impacts.

Each course was of seven-month duration and involved theoretical part and practical training (Sausen, 1991). The course was open for nationals from developing countries, nominated by their institutions. They should have graduated in natural sciences, and they could be geographer, geologist, agronomer, ecologist, biologist, forestry engineer, cartographer, oceanographer, city planner, etc. A large number of specialists have been trained in this way from Latin America (Argentina, Brazil, Cuba, Costa Rica, Chile, Colombia, Dominion Republic, Equador, Mexico, Panama, Paraguai, Peru, Uruguai, Venezuela) and Africa (Kenya, Nigeria, Etheopia, Burkina Faso, Algeria, Ghana).

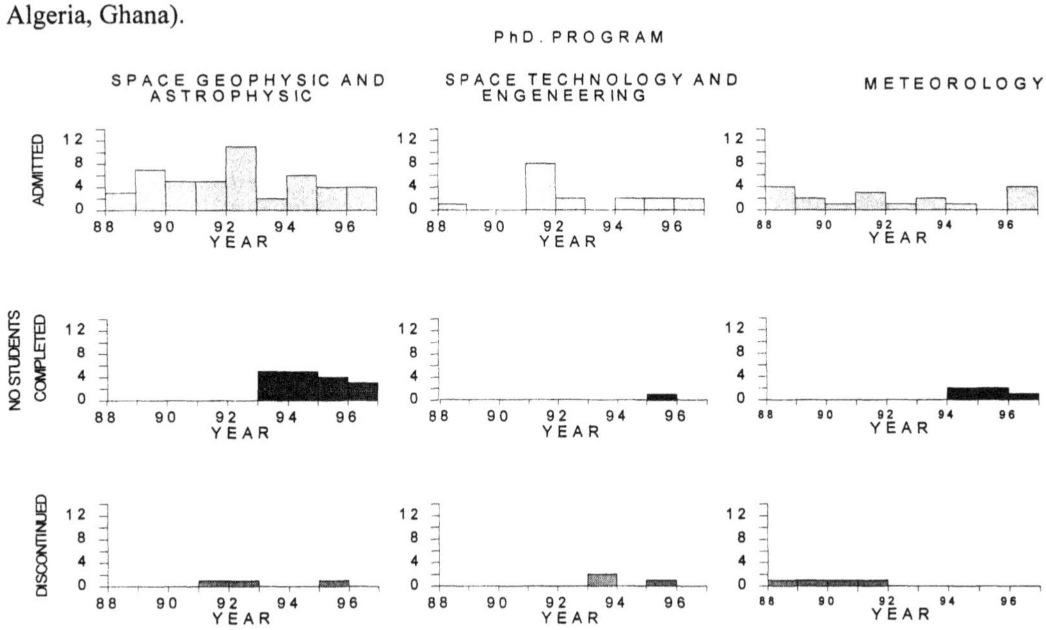

Fig. 3. Statistical representation of students admitted to PhD degree at INPE.

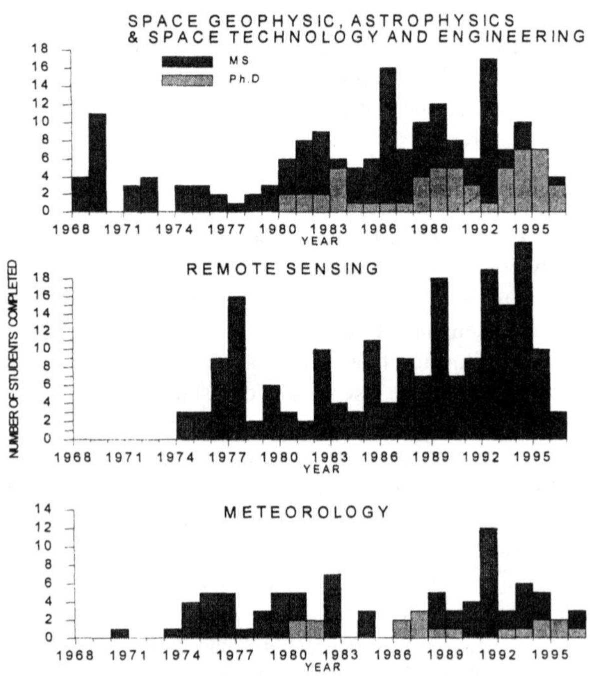

Fig. 4. Number of graduated students (Master & PhD) in each area.

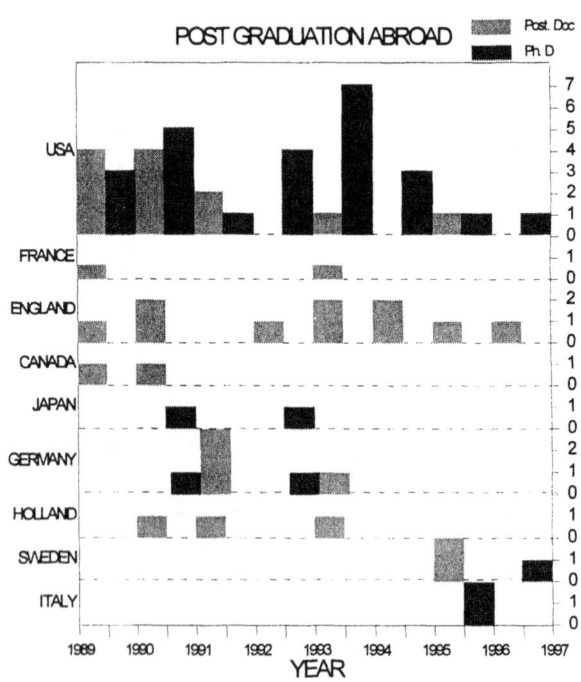

Fig. 5. Number of students who completed PhD & post doctoral research abroad.

DISCUSSION AND CONCLUSIONS

Over the years, INPE's program of space science education and training has grown significantly accompanied by large year-to-year fluctuations intake of students. Besides the programs and activities conducted by INPE, some universities in Brazil offer courses in meteorology and remote sensing. The most notable in this respect is, perhaps, the University of São Paulo where both under-graduate and post-graduate courses in meteorology are well established. A few other universities offer under-graduate courses in meteorology. Nucleus of research activities in space sciences and application areas are emerging at university centers in north (University of Northern Rio Grande, at Natal, Federal University at Campina Grande) and south (Federal University at Santa Maria, Southern Rio Grande) in Brazil. These developments were possible largly due to the initiative taken by INPE over the years to establish collaboration projetcs with the universities which involved post-graduate level training of university personnel at INPE. Also, efforts are made for information dissemination through public communication media aiming at increasing public awareness and appreciation of space science activities and the benefits therefrom.

ACKNOWLEDGEMENTS:

The authors thank Adélio Gurgel do Amaral, chief of the post-graduate service at INPE, and Tania Maria Sausen of remote sensing department (presently at the Inter Amerian Institute at INPE) for their collaboration in providing many useful information contained in this paper.

REFERENCES:

Sausen, T.M., International training course on remote sensing-Brazil: Five years training specialists in remote sensing applied to natural resources, *Report, INPE-5316-PRE/1713, 1991.*

AUTHOR INDEX

www.ingramcontent.com/pod-product-compliance
Lightning Source LLC
LaVergne TN
LVHW082353200926
848481LV00012B/3554